T0303917

Optimal Supervisory Control
of
Automated Manufacturing Systems

YuFeng Chen
Xidian University, Xi'an, China

ZhiWu Li
Xidian University, Xi'an, China

CRC Press
Taylor & Francis Group
Boca Raton London New York

CRC Press is an imprint of the
Taylor & Francis Group, an **informa** business

A SCIENCE PUBLISHERS BOOK

CRC Press
Taylor & Francis Group
6000 Broken Sound Parkway NW, Suite 300
Boca Raton, FL 33487-2742

© 2013 Copyright reserved
CRC Press is an imprint of Taylor & Francis Group, an Informa business

No claim to original U.S. Government works

Printed in the United States of America on acid-free paper

International Standard Book Number: 978-1-4665-7753-4 (Hardback)

Library of Congress Cataloging-in-Publication Data

Chen, Yufeng.
 Optimal supervisory control of automated manufacturing systems / Yufeng
   Chen, Zhiwu Li.
     p. cm.
 Includes bibliographical references and index.
 ISBN 978-1-4665-7753-4 (hardback)
 1. Manufacturing processes--Automation. 2. Production engineering--Au-
   tomation. I. Li, Zhiwu. II. Title.
 TS183.C483 2012
 670.285--dc23
                                                          2012036497

Visit the Taylor & Francis Web site at
http://www.taylorandfrancis.com

CRC Press Web site at
http://www.crcpress.com

Science Publishers Web site at
http://www.scipub.net

Preface

Discrete event systems (DESs) are an outcome of the development of computer and information technology, which in recent decades have become an integral part of our world. They encompass a wide variety of physical systems that arise from contemporary technology, such as urban transportation systems, logistic systems, database management systems, communication protocols, computer communication networks, distributed software systems, monitoring and control systems of large buildings, air and train traffic control systems, and highly integrated command, control, communication, and intelligence systems.

As a typical example of DESs, automated manufacturing systems (AMSs), usually considered to be an innovative, agile, and quick response pattern of production, have received much attention in recent decades since the traditional mass production mode is challenged by the quick changes of market requirements.

Due to a high degree of resource sharing, there exist deadlocks in an AMS, which are an undesirable phenomenon since their occurrence usually gives rise to unnecessary productivity loss and even catastrophic results in highly automated systems such as semiconductor manufacturing and safety-critical distributed databases. Deadlock problems in AMSs have received more and more attention from both academic and industrial communities.

Digraphs, automata, and Petri nets are three major mathematical tools to investigate deadlock problems in AMSs. Recent decades have seen that Petri nets are increasingly becoming an important, popular, and fully-fledged mathematical model to provide solutions to the issues. There are three criteria to evaluate and design a liveness-enforcing Petri net supervisor for an AMS to be controlled, which takes the form of monitors, sometimes called control places, that can be regarded as the intervention from human beings or other external agencies. The criteria include behavioral permissiveness, computational complexity, and structural complexity.

A maximally permissive supervisor implies that all legal states in the sense of deadlock control in a plant to be controlled are reachable in the controlled system, which, from the productivity point of view, usually leads to high utilization of system resources. A deadlock control algorithm with low computational complexity usually means that the calculation of its corresponding supervisor is tractable and that it can potentially be applied to the real-world systems. Structural complexity of a liveness-enforcing supervisor is referred to as the number of monitors as well as related arcs in the supervisor. A supervisor with a small number of monitors can always decrease the hardware and software costs in the stage of model checking and verification, and control validation and implementation.

In general, it is difficult or even impossible, given a real-world system, to find a maximally permissive, yet computationally efficient, supervisor with a minimal number of monitors. A trade-off among behavioral permissiveness, structural complexity, and computational tractability is usually adopted. For example, siphon-based deadlock prevention approaches that do not depend on a partial or complete state enumeration cannot in general lead to a maximally permissive supervisor. On the other hand, most deadlock prevention approaches, existing in the literature, that can derive maximally permissive liveness-enforcing supervisors expressed by a set of monitors depend on a complete marking enumeration except for some net subclasses at special initial markings. This monograph aims to present the state-of-the-art developments in the design of behaviorally and structurally optimal liveness-enforcing Petri net supervisors with computationally tractable approaches. The outline of this book is as follows:

Chapter 1 introduces AMSs with focus on their deadlock control issues. A brief review is provided of a variety of deadlock avoidance and prevention approaches in the literature.

Chapter 2 recalls the basic concepts and definitions of Petri nets, including siphons, P-invariants, state equations, reachability sets, and reachability graphs. Also, binary decision diagrams (BDDs) are introduced as a powerful tool to analyze Petri nets. It offers fundamentals for readers to understand the essential contributions and makes the book self-contained.

Reachability graphs are the most powerful analysis technique of Petri nets, whose computation usually suffers from the state explosion problem. Chapter 3 formulates symbolic computation and analysis methods of bounded Petri nets, by using BDDs which are capable of representing large sets of markings with small data structures. The use of symbolic computation and analysis makes it possible to compute an optimal liveness-enforcing supervisor for large-sized systems.

The theory of regions deals with the synthesis problem of Petri nets from automaton-based behavioral descriptions. Chapter 4 proceeds to a deadlock prevention strategy by using the theory of regions, which can lead to a maximally permissive liveness-enforcing supervisor expressed by a set of monitors if, given a plant, such a supervisor exists. The major issue of this strategy is the computational complexity problem.

Chapter 5 considers the design of a maximally permissive liveness-enforcing supervisor for manufacturing-oriented Petri net models existing in the literature. Once the reachability graph of a plant net model is computed, it is divided into two parts: a live-zone and a deadlock-zone. By a vector covering approach, two sets of reachable and forbidden markings, a minimal covering set of legal markings and a minimal covered set of first-met bad markings, are defined. A maximally permissive liveness-enforcing supervisor is referred to as a set of monitors such that all elements in the minimal covering set of legal markings are reachable and no element in the minimal covered set of first-met bad markings is reachable.

Motivated by the fact that a maximally permissive supervisor described by an automaton does not always admit a Petri net representation, Chapter 6 undertakes the most behaviorally permissive supervisor design problem. A liveness-enforcing Petri net supervisor is said to be the most permissive if there are no other pure Petri net supervisors more permissive than it.

The structural complexity of a liveness-enforcing Petri net supervisor is usually represented by the number of its monitors. Structurally simple supervisors imply the low computational overheads in model checking, validation, and system implementation. Chapter 7 considers the design of a maximally permissive liveness-enforcing supervisor with a compact supervisory structure by minimizing the number of monitors.

Chapter 8 provides a well trade-off among the three criteria: behavioral permissiveness, structural complexity, and computational complexity. An iterative approach is proposed to design a liveness-enforcing supervisor that is behaviorally optimal with a small number of control places. Meanwhile, the computational overhead is significantly reduced.

Chapter 9 deals with the forbidden state problems that are a typical class of control specifications in supervisory control of DESs. The chapter provides a methodology to design a maximally permissive supervisor that prevents the reachability of a given set of forbidden states only, while the supervisor is structurally minimized. As another typical class of control specifications, generalized mutual exclusion constraints (GMECs) are also considered in this chapter by presenting a maximally permissive supervisor with a minimal supervisory structure. In other words, by using a minimal number of monitors, all states satisfying the set of GMECs are reachable while the ones violating the constraints are forbidden.

Finally, Chapter 10 concludes the book and offers a number of open technical problems and future research directions.

Attached to the end of every chapter is a reference bibliography. A glossary and a complete index are provided in the final part, which should facilitate readers in using this book.

The monograph evolves from the recent research work in System Control & Automation Group, Xidian University, which is originally sparked by recognizing the technical faults of the theory of regions when, pioneered by Professor Murat Uzam, it is applied to the deadlock prevention problem for manufacturing systems.

Readers are, on the whole, expected to come to understand the state-of-the-art developments of optimal supervisory control problems arising in automated production systems. The optimality of a supervisor that is expressed by a set of monitors implies that it is maximally permissive and structurally minimal with computationally reasonable overheads. The readers can learn a methodology to achieve the optimality purposes of deadlock prevention via converting a variety of problems under consideration into integer linear programming models. The approaches reported in this book convince the readers of their significance in dealing with other supervisory control problems in DESs.

YuFeng Chen
September 2012
ZhiWu Li

Acknowledgements

We are very grateful to Professor MengChu Zhou, Department of Electrical and Computer Engineering, New Jersey Institute of Technology. Since 2002, we have been collaborating in supervisory control of automated manufacturing systems, particularly, in deadlock analysis and control issues. In 2007, Professor Zhou was invited by Xidian University as a Lecture Professor sponsored by the Cheung Kong Scholars Programme, launched by the Ministry of Education, China, and Hong Kong Li KaShing Foundation. Professor Zhou's affiliation to Xidian University leads to the birth of System Control & Automation Group, currently run by Professor ZhiWu Li, the co-author of this monograph. Our sincere thanks go to Professor Zhou since the appearance of this book is impossible without his valuable suggestions, critical comments and reviews, sweet encouragement, and kind support.

We extend very special thanks to many people who directly or indirectly contribute in a variety of ways to the development of the material included in this book. The continuing interaction and stimulating discussions with them have been a constant source of encouragement and inspiration. They include Professors M. D. Jeng, Taiwan Ocean University (China), Y. S. Huang, Taiwan ILan University (China), M. Uzam, Niğde Üniversitesi, Y. Chao, National Chengchi University (Taiwan, China), M. P. Fanti, Polytechnic di Baris, J. C. Wang, University of Monmouth, F. Lewis, University of Texas at Arlington, M. Khalgui and O. Mosbahi, University of Carthage, L. Feng, Royal Institute of Technology, F. Tricas, Universidad de Zaragoza, N. Q. Wu, GuangDong Institute of Technology, K. Y. Xing, Xi'an Jiaotong University, W. M. Wu, ZheJiang University, L. Wang, Peking University, and S. G. Wang, ZheJiang Gongshang University.

We would like to express our sincere gratitude and appreciation to Professor W. M. Wonham, Department of Electrical and Computer Engineering, University of Toronto, Professor M. Shpitalni, Department of Mechanical Engineering, Israel Institute of Technology (Technion), Professor H. M. Hanisch, Institute of Computer Science, Martin-Luther Universität, and Professor K. Barkaoui, Cédric laboratory, Conservatoire National Des Arts Et Métiers (Cnam), who hosted the second author of this book as a visiting professor in 2002, 2007, 2008, and 2010, respectively.

This monograph was in part supported by the National Natural Science Foundation of China under Grant Nos. 59505022, 60474018, 60773001, 61074035, and 61203038, the Fundamental Research Funds for the Central Universities under Grant Nos. JY10000904001, K50510040012, and K5051204002, the National

Research Foundation for the Doctoral Program of Higher Education, the Ministry of Education, P. R. China, under Grant Nos. 20070701013 and 20090203110009, the Research Fellowship for International Young Scientists, the National Natural Science Foundation of China, under Grant No. 61050110145, the Cheung Kong Scholars Programme, the Ministry of Education, P. R. China, 863 High-tech Research and Development Program of China under Grant No. 2008AA04Z109, and Alexander von Humboldt Foundation.

YuFeng Chen
ZhiWu Li

September 2012

Contents

Contents

Acronyms

AMS	automated manufacturing system
CPN	colored Petri net
DDD	data decision diagram
DES	discrete event system
DZ	deadlock-zone
ES^{3}PR	extended system of simple sequential processes with resources
FBM	first-met bad marking
FMS	flexible manufacturing system
GMEC	generalized mutual exclusion constraint
ILPP	integer linear programming problem
LES	liveness-enforcing supervisor
LPP	linear programming problem
LZ	live-zone
MCPP	minimal number of control places problem
MDD	multi-valued decision diagram
MFFP	maximal number of forbidding FBM problem
MIP	mixed integer programming
MLMP	most legal markings problem
MTSI	marking/transition separation instance
OBDD	ordered binary decision diagram
OSDTR	optimal supervisor design by the theory of regions
OSFS	optimal supervisor for forbidden states
PI	place invariant
PN	Petri net
PPN	production Petri net
RAS	resource allocation system
ROBDD	reduced ordered binary decision diagram
ROPN	resource-oriented Petri net
SCT	supervisory control theory
S^{2}LSPR	system of simple linear sequential processes with resources
S^{3}PGR^2	system of simple sequential processes with general resource requirements
S^{3}PMR	system of simple sequential processes with multiple resources
S^{3}PR	system of simple sequential processes with resources
S^{4}R	system of sequential systems with shared resources
WS^{3}PR	weighted system of simple sequential processes with resources
ZBDD	zero-suppressed binary decision diagram

Authors

YuFeng Chen received his B.S. and Ph.D. degrees both in Mechanical Engineering from Xidian University, Xi'an, China, in 2006 and 2011, respectively. Since July 2011, he has been with School of Eletro-Mechanical Engineering, Xidian University. His interests include Petri net theory and applications, and supervisory control of discrete event systems. He serves as a frequent reviewer for a number of IEEE Transactions.

ZhiWu Li received his B.S., M.S., and Ph.D. degrees in mechanical engineering, automatic control, and manufacturing engineering, respectively, from Xidian University, Xi'an, China, in 1989, 1992, and 1995, respectively. He joined Xidian University, in 1992, where he is currently the dean and a Professor of School of Electro-Mechanical Engineering and the director of Systems Control and Automation Group. From June 2002 to July 2003, he was a Visiting Professor at the Systems Control Group, Department of Electrical and Computer Engineering, University of Toronto, Toronto, ON, Canada. From February 2007 to February 2008, he was a Visiting Scientist at the Laboratory for Computer-Aided Design (CAD) & Lifecycle Engineering, Department of Mechanical Engineering, Technion-Israel Institute of Technology, Technion City, Haifa, Israel. Since November 2008, he has been a Visiting Professor in Automation Technology Laboratory, Institute of Computer Science, Martin-Luther University of Halle-Wittenberg, Halle (Saale), Germany. He visited Cédric laboratory, Conservatoire National Des Arts Et Métiers (Cnam), Paris, France, in 2010 and 2012. He serves as a host professor of Research Fellowship for International Young Scientists, National Natural Science of Foundation of China.

He is the author or coauthor of over 150 publications including a book chapter in Deadlock Resolution in Computer-Integrated Systems (Marcel Dekker, 2005). He is a coauthor with MengChu Zhou, Deadlock Resolution in Automated Manufacturing Systems: A Novel Petri Net Approach (Springer, 2009) and Modeling, Analysis and Deadlock Control in Automated Manufacturing Systems (Beijing, 2009, in Chinese). His current research interests include Petri net theory and application, supervisory control of discrete event systems, workflow modeling and analysis, and systems integration.

He is the General Co-Chair of the IEEE International Conference on Automation Science and Engineering, August 23-26, Washington, DC, 2008. He is a financial

Co-Chair of the IEEE International Conference on Networking, Sensing, and Control, March 26–29, 2009, a member of International Advisory Committee, 10th International Conference on Automation Technology, June 27–29, 2009, a Co-Chair of the program committee of the IEEE International Conference on Mechatronics and Automation, August 24-27, 2010, and members of the program committees of many international conferences. He serves an Associate Editor of the IEEE Transactions on Automation Science and Engineering, IEEE Transactions on Systems, Man, and Cybernetics, Part A: Systems and Humans, International Journal of Embedded Control Systems, IST Transactions of Robotics, Automation & Mechatronics—Theory & Applications, and IST Transactions of Control Engineering-Theory and Applications. He is a Guest Editor of Special Issue on "Petri Nets for System Control and Automation" in Asian Journal of Control, Special Issue on "Petri Nets and Agile Manufacturing" in Transactions of the Institute of Measurement and Control, and Special Issue on "Modeling and Verification of Discrete Event Systems" in ACM Transactions on Embedded Computing Systems.

He is a member of Discrete Event Systems Technical Committee of the IEEE Systems, Man, and Cybernetics Society. He serves as a frequent reviewer for more than 30 international journals including a number of IEEE Transactions as well as many international conferences. He is listed in Marquis Who's Who in the world, 27th Edition, 2010. Dr. Li is a recipient of Alexander von Humboldt Research Grant, Alexander von Humboldt Foundation, Germany. He is a senior member of IEEE and the founding chair of Xi'an Chapter of IEEE Systems, Man, and Cybernetics Society.

Chapter 1

Introduction

ABSTRACT

This chapter outlines deadlock resolution by Petri nets in automated manufacturing systems. Interesting work for deadlock control of automated manufacturing systems is briefly reviewed. Also, both advantages and disadvantages of the previous work are provided to show the development of the deadlock control approaches by using Petri nets.

1.1 AUTOMATED MANUFACTURING SYSTEMS

An automated manufacturing system (AMS) is a computer-controlled system which can work twenty-four hours a day. It can automatically finish various kinds of jobs by using shared resources such as robots, machines, and automated guided vehicles. In an AMS, raw parts are processed concurrently in a pre-established sequence to compete for limited system resources. The competition may cause deadlocks (Coffman *et al.*, 1971; Gligor and Shattuck, 1980) when some processes keep waiting indefinitely for the other processes to release resources. Deadlocks are a highly undesirable situation that must be considered in AMSs since a system in a deadlock situation always means that the whole system or a part of it is blocked. Deadlocks often offset the advantages of these systems. For example, deadlocks in a flexible manufacturing system (FMS) can cause unnecessary cost, such as long downtime and low use of some critical and expensive resources, and may lead to catastrophic results. Therefore, many researchers do more and more work to deal with deadlock problems in AMSs.

Coffman *et al.* (1971) establish four necessary conditions for the occurrence of deadlocks in a resource allocation system (RAS), which can be explained in the context of AMSs as follows:

(1) "Mutual exclusion", which means that at any time, a resource can only be exclusively occupied by one part. For other parts, this resource is unavailable until it is released;

(2) "No preemption", which implies that once a resource is acquired by a part in process, it cannot be forcibly removed by any external agent. The resource can be released only by the explicit action of the process holding it;

(3) "Hold and wait", which indicates that processes already hold some and wait for additional resources;

(4) "Circular wait", which infers that there is a set of linearly ordering processes such that each process requests the resources currently held by the next process while the last process requests the resources held by the first.

Once a deadlock occurs, the above four conditions must be held. That is to say, deadlocks cannot occur if one of them is broken. The first three conditions depend on the physical property of a system and its resources. However, the last condition is decided by the request, allocation, and release of system resources. It is controllable and can be broken by properly assigning the resources of a system, aiming to avoid the occurrence of a circular wait.

Generally, deadlock resolution methods are classified into three strategies: deadlock detection and recovery (Kumaran et al., 1994; Wysk et al., 1994), deadlock avoidance (Abdallah and ElMaraghy, 1998; Banaszak and Krogh, 1990; Ezpeleta et al., 2002, 2004; Hsieh and Chang, 1994; Hsieh, 2004; Park and Reveliotis, 2000; Viswanadham et al., 1990; Wu, 1999; Wu and Zhou, 2001, 2005; Xing et al., 1995, 1996), and deadlock prevention (Ezpeleta et al., 1995; Fanti and Zhou, 2004, 2005; Jeng and Xie, 2005; Lautenbach et al., 1996; Tricas et al., 1998, 2000).

A deadlock detection and recovery approach is a straight way to handle the deadlock problems. It does not attempt to prevent deadlocks. Instead, it allows the occurrences of deadlocks. Once a deadlock occurs and is detected, some actions are taken to reallocate the resources to put back the system to a deadlock-free state. The approach needs to detect the deadlocks periodically, which requires a large amount of data and may become complex if several types of shared resources are considered (Abdallah and ElMaraghy, 1998). The efficiency of this approach depends upon the response time of the implemented algorithms for deadlock detection and recovery. Meanwhile, it is always applied to the systems where recovery strategies are easy to implement.

A deadlock avoidance approach intends to be easy to implement. It dynamically examines each system state to ensure that deadlocks can never exist. At each state, a proper algorithm is implemented to make a correct decision to proceed among all feasible evolutions, which can keep a system away from deadlock states. It is an aggressive method since it tries to allow more states of a system, even sometimes deadlocks cannot be totally eliminated. Hence, it usually leads to high resource utilization and throughput. However, once it cannot avoid all deadlocks, suitable recovery strategies are still required (Kumaran et al., 1994; Viswanadham et al., 1990; Wysk et al., 1994).

A deadlock prevention approach always uses an off-line computational mechanism to impose constraints on a system to prevent it from reaching deadlock states. The request for resources is controlled by a proper strategy to exclude the happening of deadlocks. Its advantage is that once a deadlock control policy is established, deadlocks can never occur. Compared with deadlock avoidance, deadlock prevention tends to be too conservative, which reduces the resource utilization and system productivity. However, it has the advantages of safety and stability.

Generally, deadlocks in AMSs can be handled by a number of formal tools, such as graph theory, automata, and Petri nets. Graph theory or a digraph is a simple and intuitive tool to describe interactions between operations and resources, from which a deadlock control policy can be derived. In the framework of graph theory, deadlocks are always related with the circuits of the graph. Thus, deadlock occurrences in an AMS can be detected by easily computing all the circuits. Such an algorithm has been deeply investigated in the field of graph theory and many theoretical results can be applied to deadlock resolution. The representative work can be found in the literature by Wysk (Cho *et al.*, 1995; Kumaran *et al.*, 1994; Wysk *et al.*, 1991, 1994), and Fanti (Fanti *et al.*, 1996a,b, 1997, 1998, 2000, 2001; Fanti, 2002; Fanti *et al.*, 2002). Originated by Ramadge and Wonham, supervisory control theory (SCT) (Ramadge and Wonham, 1987, 1989) based on formal languages and finite automata aims at providing a comprehensive and structural treatment of the modeling and control of discrete event systems (DESs). As an important paradigm, SCT has a profound influence on the supervisory control of AMSs under other formalisms such as Petri nets. A number of effective yet computationally efficient deadlock control policies are developed based on automata. A large amount of representative research is presented by Lawley, Reveliotis, and Ferreira in (Lawley *et al.*, 1997a,b,c, 1998a,b, 1999; Lawley, 2000; Lawley and Reveliotis, 2001; Lawley and Sulistyono, 2002; Reveliotis and Ferreira, 1996; Reveliotis *et al.*, 1997; Reveliotis, 2007; Yalcin and Boucher, 2000). In particular, a theoretically significant deadlock avoidance policy with polynomial complexity is developed for a class of resource allocation systems in (Reveliotis *et al.*, 1997), which is then described in a Petri net formalism (Park and Reveliotis, 2001).

1.2 SUPERVISORY CONTROL OF AUTOMATED MANUFACTURING SYSTEMS

Petri nets (Murata, 1989) are a graph-based mathematical formalism suitable to describe, model, and analyze the behavior of RASs. As a good tool for describing and analyzing AMSs, Petri nets can reflect their behavior and properties, including the orders of events, concurrency, synchronization, and deadlocks. Now Petri nets have been used in a wide range of areas such as manufacturing systems, computer and communication networks, and automation systems.

Based on Petri nets, researchers develop many policies to deal with the deadlock problems (Ezpeleta *et al.*, 1995; Li and Zhou, 2004, 2005, 2006b; Li *et al.*, 2007, 2008a). Generally, there are mainly two analysis techniques to deal with deadlock prevention in AMSs: structure (Huang *et al.*, 2001, 2006; Jeng and Xie, 2005; Kumaran *et al.*, 1994; Li and Zhou, 2006a,b) and reachability graph (Ghaffari *et al.*, 2003; Uzam, 2002) analysis. The former always obtains a deadlock prevention policy by special structural objects of a Petri net such as siphons and resource-transition circuits. This method can lead to a computationally efficient liveness-enforcing supervisor in general but always restricts a system such that a portion of permissive behavior is excluded. For the latter, the reachability graph can fully reflect the behavior of a system. Though its computation is very expensive, a rather highly or even maximally permissive liveness-enforcing supervisor can always be obtained.

The above two analysis techniques play a key role in the development of deadlock control policies of Petri nets. A very popular software package for analysis of Petri nets is INA (Starke, 2003). However, it cannot handle large-scale Petri net systems, especially in the sense of reachability graph analysis. For instance, based on a computer in the Windows XP operating system with Intel CPU Core 2.8 GHz and 4 GB memory, we have tried to enumerate the reachable markings for an example with 48 places and 38 transitions by using INA. However, the computation cannot be finished due to the memory overflow problem. Another powerful tool for generating reachable markings of a Petri net is binary decision diagrams (BDDs) (Andersen, 1997; Brant, 1992). BDDs have a compact data structure that makes them have the capability of representing large sets of encoded data with a small data structure and enable the efficient manipulation of those sets. BDDs have been applied to the analysis of Petri nets successfully to deal with some problems. For example, in (Pastor *et al.*, 1994, 2001; Miner and Ciardo, 1999), they are used to model the structure and behavior of bounded Petri nets, providing a very efficient way to compute the set of reachable markings. This implies that BDDs are powerful to deal with large sets of data.

There are three important criteria in evaluating the performance of a liveness-enforcing supervisor for a system to be controlled, which are behavioral permissiveness, structural complexity, and computational complexity. A maximally permissive supervisor always implies high utilization of system resources. A supervisor with a small number of control places can decrease the hardware and software costs in the stage of control verification, validation, and implementation. A deadlock control policy with low computational complexity means that it can be applied to large-sized systems. Thus, many efforts are made to develop deadlock prevention algorithms that can obtain liveness-enforcing supervisors with maximal permissiveness, simple supervisory structures, and low computational costs.

Reachability graph analysis is an important technique for deadlock control, which always suffers from a state explosion problem since it requires generating all or a part of reachable markings. Based on this technique, an optimal or suboptimal supervisor with highly behavioral permissiveness can always be achieved. Uzam and Zhou (2006) develop an iterative approach to design an optimal or suboptimal supervisor. In their study, the reachability graph of a net is classified into two parts: a live-zone (LZ) and a deadlock-zone (DZ). First-met bad markings (FBMs) are derived from the reachability graph. An FBM is a marking in the DZ, representing the very first entry from the LZ to the DZ. At each iteration, an FBM is selected and a control place is designed to prevent the FBM from being reached by using a place invariant (PI) based method proposed in (Yamalidou *et al.*, 1996). This process does not terminate until the resulting net is live. This method is easy to use if the reachable space of a system is small but cannot guarantee the behavioral optimality of the supervisor.

In (Ghaffari *et al.*, 2003), the theory of regions is used as an effective approach since it can definitely find an optimal supervisor if it exists. However, it suffers from computational and structural complexity problems. By combining siphons and marking generation, Piroddi *et al.* (2008) propose a selective siphon control policy that can obtain a small-sized supervisor with highly permissive behavior. A modified approach is provided in (Piroddi *et al.*, 2009) to avoid a full siphon enumeration that makes more efficient the deadlock prevention policy in (Piroddi *et al.*, 2008). The two methods can find an optimal supervisor for each example presented in their studies (Piroddi *et al.*, 2008, 2009). However, no formal proof is provided to show that their policy is definitely maximally permissive in theory. On the other hand, they reduce the complexity of supervisory structures but cannot minimize them.

1.3 SUMMARY

Behavioral permissiveness plays an important role in the development of deadlock control of AMSs. Many researchers and engineers have done a lot of work to deal with the deadlock problems in AMSs and try to make the controlled systems with as many admissible states as possible. In this particular monograph, we focus on the optimal behavioral permissiveness of Petri net supervisors. Meanwhile, both structural complexity and the computational complexity are considered. Most of the new methods for the design of optimal Petri net supervisors are included in this monograph, such as the theory of regions (Ghaffari *et al.*, 2003) and some of our recent research results (Chen *et al.*, 2011, 2012; Chen and Li, 2011)

1.4 BIBLIOGRAPHICAL REMARKS

There are survey papers and books dealing with the deadlocks in AMSs. The paper (Fanti and Zhou, 2004) surveys the deadlock control approaches in the literature. A lot of Petri nets based deadlock control policies are reviewed and compared in (Li *et al.*, 2008b). Also, a recent work (Li *et al.*, 2012) reviews a lot of deadlock control policies and provides some comparisons in the sense of behavioral permissiveness, structural complexity, and computational complexity. The books in the area of Petri nets for deadlock problems can be found in (Hruz and Zhou, 2007; Li and Zhou, 2009; Wu and Zhou, 2010; Zhou and DiCesare, 1993; Zhou and Venkatesh, 1998).

References

Abdallah, I. B. and H. A. ElMaraghy. 1998. Deadlock prevention and avoidance in FMS: A Petri net based approach. International Journal of Advanced Manufacturing Technology. 14(10): 704–715.

Andersen, H. R. 1997. An introduction to binary decision diagrams. Lecture Notes for 49285 Advanced Algorithms E97, Department of Information Technology, Technical University of Denmark.

Banaszak, Z. and B. H. Krogh. 1990. Deadlock avoidance in flexible manufacturing systems with concurrently competing process flows. IEEE Transactions on Robotics and Automation. 6(6): 724–734.

Brant, R. 1992. Symbolic Boolean manipulation with ordered binary decision diagrams. ACM Computing Surveys. 24(3): 293–318.

Chen, Y. F. and Z. W. Li. 2011. Design of a maximally permissive liveness-enforcing supervisor with a compressed supervisory structure for flexible manufacturing systems. Automatica. 47(5): 1028–1034.

Chen, Y. F., Z. W. Li, M. Khalgui, and O. Mosbahi. 2011. Design of a maximally permissive liveness-enforcing Petri net supervisor for flexible manufacturing systems. IEEE Transactions on Automation Science and Engineering. 8(2): 374–393.

Chen, Y. F., Z. W. Li, and M. C. Zhou. 2012. Behaviorally optimal and structurally simple liveness-enforcing supervisors of flexible manufacturing systems. IEEE Transactions on Systems, Man, and Cybernetics, Part A. 42(3): 615–629.

Cho, H., T. K. Kumaran, and R. A. Wysk. 1995. Graph-theoretic deadlock detection and resolution for flexible manufacturing systems. IEEE Transactions on Robotics and Automation. 11(3): 413–421.

Coffman, E. G., M. J. Elphick, and A. Shoshani. 1971. System deadlocks. ACM Computing Surveys. 3(2): 67–78.

Ezpeleta, J., J. M. Colom, and J. Martinez. 1995. A Petri net based deadlock prevention policy for flexible manufacturing systems. IEEE Transactions on Robotics and Automation. 11(2): 173–184.

Ezpeleta, J., F. Tricas, F. Garcia-Valles, and J. M. Colom. 2002. A banker's solution for deadlock avoidance in FMS with flexible routing and multiresource states. IEEE Transactions on Robotics and Automation. 18(4): 621–625.

Ezpeleta, J. and L. Recalde. 2004. A deadlock avoidance approach for nonsequential resource allocation systems. IEEE Transactions on Systems, Man, and Cybernetics, Part A. 34(1): 93–101.

Fanti, M. P., G. Maione, and B. Turchiano. 1996a. Digraph-theoretic approach for deadlock detection and recovery in flexible production systems. Studies in Informatics and Control. 5(4): 373–383.

Fanti, M. P., B. Maione, S. Mascolo, and B. Turchiano. 1996b. Performance of deadlock avoidance algorithms in flexible manufacturing systems. Journal of Manufacturing Systems. 15(3): 164–178.

Fanti, M. P., B. Maione, S. Mascolo, and B. Turchiano. 1997. Event-based feedback control for deadlock avoidance in flexible production systems. IEEE Transactions on Robotics and Automation. 13(3): 347–363.

Fanti, M. P., B. Maione, and B. Turchiano. 1998. Event control for deadlock avoidance in production systems with multiple capacity resources. Studies Informatics and Control. 7(4): 343–364.

Fanti, M. P., B. Maione, and B. Turchiano. 2000. Comparing digraph and Petri net approaches to deadlock avoidance in FMS. IEEE Transactions on Systems, Man and Cybernetics, Part B. 30(5): 783–798.

Fanti, M. P., G. Maione, and B. Turchiano. 2001. Distributed event-control for deadlock avoidance in automated manufacturing systems. International Journal of Production Research. 39(9): 1993–2021.

Fanti, M. P. 2002. Event-based controller to avoid deadlock and collisions in zone control AGVs. International Journal of Production Research. 40(6): 1453–1478.

Fanti, M. P., G. Maione, and B. Turchiano. 2002. Design of supervisors to avoid deadlock in flexible assembly systems. International Journal of Flexible Manufacturing Systems. 14(2): 157–175.

Fanti, M. P. and M. C. Zhou. 2004. Deadlock control methods in automated manufacturing systems. IEEE Transactions on Systems, Man, and Cybernetics, Part A. 34(1): 5–22.

Fanti, M. P. and M. C. Zhou. Deadlock control methods in automated manufacturing systems. pp. 1–22. In M. C. Zhou and M. P. Fanti. [eds.]. 2005. Deadlock Resolution in Computer-Integrated Systems, New York: Marcel-Dekker Co.

Ghaffari, A., N. Rezg, and X. L. Xie. 2003. Design of a live and maximally permissive Petri net controller using the theory of regions. IEEE Transactions on Robotics and Automation. 19(1): 137–142.

Gligor, V. and S. Shattuck. 1980. On deadlock detection in distributed systems. IEEE Transactions on Software Engineering. 7(3): 320–336.

Hruz, B. and M. C. Zhou. 2007. Modeling and Control of Discrete Event Dynamic Systems. Springer, London, UK.

Hsieh, F. S. and S. C. Chang. 1994. Dispatching-driven deadlock avoidance controller synthesis for flexible manufacturing systems. IEEE Transactions on Robotics and Automation. 10(2): 196–209.

Hsieh, F. S. 2004. Fault-tolerant deadlock avoidance algorithm for assembly processes. IEEE Transactions Systems, Man, and Cybernetics, Part A. 34(1): 65–79.

Huang, Y. S., M. D. Jeng, X. L. Xie, and S. L. Chung. 2001. Deadlock prevention based on Petri nets and siphons. International Journal of Production Research. 39(2): 283–305.

Huang, Y. S., M. D. Jeng, X. L.Xie, and D. H. Chung. 2006. Siphon-based deadlock prevention for flexible manufacturing systems. IEEE Transactions on Systems, Man, and Cybernetics, Part A. 36(6): 1248–1256.

Jeng, M. D. and X. L. Xie. Deadlock detection and prevention of automated manufacturing systems using Petri nets and siphons. pp. 233–281. In M. C. Zhou and M. P. Fanti. [eds.]. 2005. Deadlock Resolution in Computer-Integrated Systems. New York: Marcel-Dekker Co.

Kumaran, T. K., W. Chang, H. Cho, and A. Wysk. 1994. A structured approach to deadlock detection, avoidance and resolution in flexible manufacturing systems. International Journal of Production Research. 32(10): 2361–2379.

Lautenbach, K. and H. Ridder. 1996. The linear algebra of deadlock avoidance—A Petri net approach. No. 25–1996, Technical Report, Institute of Software Technology, University of Koblenz-Landau, Koblenz, Germany.

Lawley, M. A., S. A. Reveliotis, and P. M. Ferreira. 1997a. Design guidelines for deadlock-handling strategies in flexible manufacturing systems. International Journal of Flexible Manufacturing Systems. 9(1): 5–29.

Lawley, M. A., S. A. Reveliotis, and P. M. Ferreira. 1997b. Flexible manufacturing system structural control and the Neighborhood Policy. part 1: Correctness and scalability. IIE Transactions. 29(10): 877–887.

Lawley, M. A., S. A. Reveliotis, and P. M. Ferreira. 1997c. Flexible manufacturing system structural control and the Neighborhood Policy, part 2: Generalization, optimization, and efficiency. IIE Transactions. 29(10): 889–899.

Lawley, M. A., S. A. Reveliotis, P. M. Ferreira. 1998a. A correct and scalable deadlock avoidance policy for flexible manufacturing systems. IEEE Transactions on Robotics and Automation. 14(5): 796–809.

Lawley, M. A., S. A. Reveliotis, and P. M. Ferreira. 1998b. The application and evaluation of banker's algorithm for deadlock-free buffer space allocation in flexible manufacturing systems. International Journal of Flexible Manufacturing Systems. 10(1): 73–100.

Lawley, M. A. 1999. Deadlock avoidance for production systems with flexible routing. IEEE Transactions on Robotics and Automation. 15(3): 497–509.

Lawley, M. A. 2000. Integrating flexible routing and algebraic deadlock avoidance policies in automated manufacturing systems. International Journal of Production Research. 38(13): 2931–2950.

Lawley, M. A. and S. A. Reveliotis. 2001. Deadlock avoidance for sequential resource allocation systems: Hard and easy cases. International Journal of Flexible Manufacturing Systems. 13(4): 385–404.

Lawley, M. A. and W. Sulistyono. 2002. Robust supervisory control policies for manufacturing systems with unreliable resources. IEEE Transactions on Robotics and Automation. 18(3): 346–359.

Li, Z. W., H. S. Hu, and A. R. Wang. 2007. Design of liveness-enforcing supervisors for flexible manufacturing systems using Petri nets. IEEE Transactions on Systems, Man, and Cybernetics, Part C. 37(4): 517–526.

Li, Z. W. and M. C. Zhou. 2004. Elementary siphons of Petri nets and their application to deadlock prevention in flexible manufacturing systems. IEEE Transactions on Systems, Man, and Cybernetics, Part A. 34(1): 38–51.

Li, Z. W. and M. C. Zhou. 2005. Comparison of two deadlock prevention methods for different-size flexible manufacturing systems. International Journal of Intelligent Control Systems. 10(3): 235–243.

Li, Z. W. and M. C. Zhou. 2006a. Clarifications on the definitions of elementary siphons of Petri nets. IEEE Transactions on Systems, Man, and Cybernetics, Part A. 36(6): 1227–1229.

Li, Z. W. and M. C. Zhou. 2006b. Two-stage method for synthesizing liveness-enforcing supervisors for flexible manufacturing systems using Petri nets. IEEE Transactions on Industrial Informatics. 2(4): 313–325.

Li, Z. W., M. C. Zhou, and M. D. Jeng. 2008a. A maximally permissive deadlock prevention policy for FMS based on Petri net siphon control and the theory of regions. IEEE Transactions on Automation Science and Engineering. 5(1): 182–188.

Li, Z. W., M. C. Zhou, and N. Q. Wu. 2008b. A survey and comparison of Petri net-based deadlock prevention policy for flexible manufacturing systems. IEEE Transactions on Systems, Man, and Cybernetics, Part C. 38(2): 173–188.

Li, Z. W. and M. C. Zhou. 2009. Deadlock Resolution in Automated Manufacturing Systems: A Novel Petri Net Approach. Springer, London, UK.

Li, Z. W., M. C. Zhou, and N. Q. Wu. 2012. Deadlock control for automated manufacturing systems based on Petri nets—A literature review. IEEE Transactions on Systems, Man, and Cybernetics, Part C. 42(4): 437–462.

Miner, A. S. and G. Ciardo. 1999. Efficient reachability set generation and storage using decision diagrams. Lecture Notes in Computer Science. 1639: 6–25.

Murata, T. 1989. Petri nets: Properties, analysis and application. Proceedings of the IEEE. 77(4): 541–580.

Park, J. and S. A. Reveliotis. 2000. Algebraic synthesis of efficient deadlock avoidance policies for sequential resource allocation systems. IEEE Transactions on Robotics and Automation. 16(2): 190–195.

Park, J. and S. A. Reveliotis. 2001. Deadlock avoidance in sequential resource allocation systems with multiple resource acquisitions and flexible routings. IEEE Transactions on Automatic Control. 46(10): 1572–1583.

Pastor, E., O. Roig, J. Cortadella, and R. M. Badia. 1994. Petri net analysis using Boolean manipulation. Lecture Notes in Computer Science. 815: 416–435.

Pastor, E., J. Cortadella, and O. Roig. 2001. Symbolic analysis of bounded Petri nets. IEEE Transactions on Computers. 50(5): 432–448.

Piroddi, L., R. Cordone, and I. Fumagalli. 2008. Selective siphon control for deadlock prevention in Petri nets. IEEE Transactions on Systems, Man, and Cybernetics, Part A. 38(6): 1337–1348.

Piroddi, L., R. Cordone, and I. Fumagalli. 2009. Combined siphon and marking generation for deadlock prevention in Petri nets. IEEE Transactions on Systems, Man and Cybernetics, Part A. 39(3): 650–661.

Ramadge, P. and W. M. Wonham. 1987. Supervisory control of a class of discrete event processes. SIAM Journal on Control and Optimization. 25(1): 206–230.

Ramadge, P. and W. M. Wonham. 1989. The control of discrete event systems. Proceedings of the IEEE. 77(1): 81–89.

Reveliotis, S. A. and P. M. Ferreira. 1996. Deadlock avoidance policies for automated manufacturing cells. IEEE Transactions on Robotics and Automation. 12(6): 845–857.

Reveliotis, S. A., M. A. Lawley, and P. M. Ferreira. 1997. Polynomial-complexity deadlock avoidance policies for sequential resource allocation systems. IEEE Transactions on Automatic Control. 42(10): 1344–1357.

Reveliotis, S. A. 2007. Implicit siphon control and its role in the liveness-enforcing supervision of sequential resource allocation systems. IEEE Transactions on Systems, Man, and Cybernetics, Part A. 37(3): 319–328.

Starke, P. H. 2003. INA: Integrated Net Analyzer. http://www2.informatik.huberlin.de/starke/ina.html.

Tricas, F., F. Garcia-Valles, J. M. Colom, and J. Ezpelata. 1998. A structural approach to the problem of deadlock prevention in processes with resources, pp. 273–278. In Proceedings of WODES'98, Italy, 26–28 August.

Tricas, F., F. Garcia-Valles, J. M. Colom, and J. Ezpelata. An iterative method for deadlock prevention in FMS. pp. 139–148. In G. Stremersch [ed.]. 2000. Discrete Event Systems: Analysis and Control, Kluwer Academic, Boston MA.

Uzam, M. 2002. An optimal deadlock prevention policy for flexible manufacturing systems using Petri net models with resources and the theory of regions. International Journal of Advanced Manufacturing Technology. 19(3): 192–208.

Uzam, M. and M. C. Zhou. 2006. An improved iterative synthesis method for liveness enforcing supervisors of flexible manufacturing systems. International Journal of Production Research. 44(10): 1987–2030.

Viswanadham, N., Y. Narahari, and T. Johnson. 1990. Deadlock prevention and deadlock avoidance in flexible manufacturing systems using Petri net models. IEEE Transactions on Robotics and Automation. 6(6): 713–723.

Wu, N. Q. 1999. Necessary and sufficient conditions for deadlock-free operation in flexible manufacturing systems using a colored Petri net model. IEEE Transactions on Systems, Man, and Cybernetics, Part C. 29(2): 192–204.

Wu, N. Q. and M. C. Zhou. 2001. Avoiding deadlock and reducing starvation and blocking in automated manufacturing systems. IEEE Transactions on Robotics and Automation. 17(5): 658–669.

Wu, N. Q. and M. C. Zhou. 2005. Modeling and deadlock avoidance of automated manufacturing systems with multiple automated guided vehicles. IEEE Transactions on Systems, Man, and Cybernetics, Part B. 35(6): 1193–1202.

Wu, N. Q. and M. C. Zhou. 2010. System Modeling and Control with Resource-Oriented Petri Nets. CRC Press, New York.

Wysk, R. A., N. S. Yang, and S. Joshi. 1991. Detection of deadlocks in flexible manufacturing cells. IEEE Transactions on Robotics and Automation. 7(6): 853–859.

Wysk, R. A., N. S. Yang, and S. Joshi. 1994. Resolution of deadlocks in flexible manufacturing systems: Avoidance and recovery approaches. Journal of Manufacturing Systems. 13(2): 128–138.

Xing, K. Y., B. S. Hu, and H. X. Chen. Deadlock avoidance policy for flexible manufacturing systems. pp. 239–263. In M. C. Zhou [ed]. 1995. Petri Nets in Flexible and Agile Automation, Kluwer Academic, Boston MA.

Xing, K. Y., B. S. Hu, and H. X. Chen. 1996. Deadlock avoidance policy for Petri-net modeling of flexible manufacturing systems with shared resources. IEEE Transactions on Automatic Control. 41(2): 289–295.

Yalcin, A. and T. O. Boucher. 2000. Deadlock avoidance in flexible manufacturing systems using finite automata. IEEE Transactions on Robotics and Automation. 16(4): 424–429.

Yamalidou, K. J., Moody, M. Lemmon, and P. Antsaklis. 1996. Feedback control of Petri nets based on place invariants. Automatica. 32(1): 15–28.

Zhou, M. C. and F. DiCesare. 1993. Petri Net Synthesis for Discrete Event Control of Manufacturing Systems. Kluwer Academic Publishers, London, UK.

Zhou, M. C. and K. Venkatesh. 1998. Modeling, Simulation and Control of Flexible Manufacturing Systems: A Petri Net Approach. World Scientific, Singapore.

Chapter 2

Preliminaries

ABSTRACT

This chapter reviews basic concepts of Petri nets, including their formal definitions, structural analysis, and reachability graph analysis used throughout this book. The definitions of Boolean algebra and binary decision diagrams (BDDs) are also reported, which are rather efficient to compute reachable markings of a Petri net model. More details can be found in (Murata, 1989; Li and Zhou, 2009; Wu and Zhou, 2010) for Petri nets and (Andersen, 1997; Pastor *et al.*, 1994, 2001) for BDDs.

2.1 INTRODUCTION

Petri nets (Wikipedia, online) were introduced by C. A. Petri in 1962 in his doctoral dissertation for the purposes of describing chemical processes. Since then, Petri nets have been widely studied and developed by researchers and engineers in manufacturing and automation areas (Abdallah and ElMaraghy, 1998; Chen and Li, 2011; Chen *et al.*, 2011; Ezpeleta *et al.*, 1995; Ghaffari *et al.*, 2003; Hu *et al.*, 2010; Li *et al.*, 2007; Piroddi *et al.*, 2008). Petri nets can model concurrent systems and their behavior with a mathematical formalism.

Compared with Turing machines and automata (Ramadge and Wonham, 1987, 1989), Petri nets have less modeling and decision power. Turing machines are of complete expressiveness for any hardware or software issues. Even automata, they have more modeling power for a system with an infinite number of states. However, Petri nets still have been widely used to model, analyze, and control AMSs thanks to their property that is suitable to detect deadlocks (Coffman *et al.*, 1971) in these systems and establish a policy to deal with deadlocks (Fanti *et al.*, 2000; Fanti and Zhou, 2004, 2005; Zhou and DiCesare, 1991; Zhou *et al.*, 1992; Zhou and DiCesare, 1993). In fact, researchers and engineers develop a lot of software packages that make Petri nets easy to be used as a systemic tool to model and handle the control of a system in the real world. This chapter provides a formal treatment of Petri nets to facilitate the understanding of the following chapters.

Siphons and reachability graphs play very important roles in the deadlock control by Petri nets (Huang *et al.*, 2001, 2006; Jeng and Xie, 2005; Li and Zhou, 2004; Piroddi *et al.*, 2009; Reveliotis, 2007). However, both increase exponentially with respect to the size of a Petri net. Deciding how to efficiently find all siphons, in fact minimal siphons, and the reachability graph of a Petri net model to be controlled, is an important issue (Miner and Ciardo, 1999). BDDs are known as a powerful tool to save and process a large set of encoded data. Thus, in this chapter, basic concepts of BDDs are also provided, which are used in Chapter 3.

2.2 PETRI NETS

2.2.1 Basic Concepts

A Petri net is a directed bipartite graph that consists of two components—a net structure and an initial marking. A net (structure) contains two types of nodes— places and transitions. There are directed arcs from places to transitions and directed arcs from transitions to places in a net. Arcs are labeled by positive integers to represent their weights and labels for a unity weight are always omitted. In a graphical representation, places are shown as circles and transitions as boxes or bars. A place can hold tokens graphically denoted by black dots, or a positive integer representing their number. The distribution of tokens over the places of a net is called a marking that corresponds to a state of the modeled system. The initial token distribution is hence called the initial marking. Let $\mathbb{N}$ denote the set of non-negative integers and $\mathbb{N}^+$ the set of positive integers.

Definition 2.1   A generalized Petri net (structure) is a four-tuple $N = (P, T, F, W)$ where P and T are finite and nonempty sets. P is a set of places and T is a set of transitions with $P \cup T \neq \emptyset$ and $P \cap T = \emptyset$. $F \subseteq (P \times T) \cup (T \times P)$ is called a flow relation of the net, represented by arcs with arrows from places to transitions or from transitions to places. $W : (P \times T) \cup (T \times P) \rightarrow \mathbb{N}$ is a mapping that assigns a weight to an arc: $W(x,y) > 0$ if $(x,y) \in F$, and $W(x,y) = 0$ otherwise, where $x,y \in P \cup T$.

EXAMPLE 2.1   A simple Petri net is shown in Fig. 2.1, where $P = \{p_1 - p_5\}$, $T = \{t_1 - t_4\}$, $F = \{(p_1,t_1), (t_1,p_2), (p_2,t_4), (t_4,p_1), (t_1,p_3), (p_3,t_2), (t_2,p_4), (p_4,t_3), (p_3,t_3), (t_3,p_5), (p_5,t_4)\}$, $W(p_1,t_1) = W(t_4,p_1) = W(t_1,p_3) = W(p_3,t_2) = W(t_2,p_4) = W(p_4,t_3) = W(p_3,t_3) = W(t_3,p_5) = W(p_5,t_4) = 1$, and $W(t_1,p_2) = W(p_2,t_4) = 2$. Places and transitions are graphically represented by circles and boxes, respectively.

Definition 2.2   A Petri net $N = (P, T, F, W)$ is said to be ordinary if $\forall (x,y) \in F$, $W(x,y) = 1$. N is said to be generalized if $\exists (x,y) \in F$, $W(x,y) > 1$.

EXAMPLE 2.2   The Petri net shown in Fig. 2.1 is generalized since $W(t_1,p_2) = W(p_2,t_4) = 2$. An ordinary net is shown in Fig. 2.2 since the weight of each arc is equal to one.

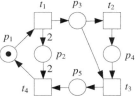

Figure 2.1 A Petri net example.

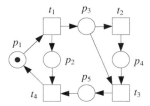

Figure 2.2 An ordinary Petri net.

Definition 2.3 A marking M of a Petri net N is a mapping from P to $\mathbb{N}$. $M(p)$ denotes the number of tokens in place p. A place p is marked at a marking M if $M(p) > 0$. A subset $S \subseteq P$ is marked at M if at least one place in S is marked at M. The sum of tokens of all places in S is denoted by $M(S)$, i.e., $M(S) = \sum_{p \in S} M(p)$. S is said to be empty at M if $M(S) = 0$. (N, M_0) is called a net system or marked net and M_0 is called an initial marking of N.

Markings and vectors are usually represented via using a multiset (bag) or formal sum notation for the sake of an expedient description. As a result, vector M is denoted by $\sum_{p \in P} M(p)p$. For instance, a marking that puts two tokens in place p_1 and three tokens in place p_4 only in a net with $P = \{p_1 - p_6\}$ is denoted as $2p_1 + 3p_4$ instead of $(2\ 0\ 0\ 3\ 0\ 0)^T$.

EXAMPLE 2.3 For the Petri net model in Fig. 2.1, its initial marking is $(1\ 0\ 0\ 0\ 0)^T$, expediently denoted as $M_0 = p_1$. At M_0, only p_1 is marked since $M_0(p_1) = 1 > 0$.

Definition 2.4 The preset (postset) of a node $x \in P \cup T$ is defined as $^\bullet x = \{y \in P \cup T | (y, x) \in F\}$ $(x^\bullet = \{y \in P \cup T | (x, y) \in F\})$.

Note that the preset (postset) of a set $X \subseteq P \cup T$ can be accordingly defined as $^\bullet X = \cup_{x \in X} {}^\bullet x$ $(X^\bullet = \cup_{x \in X} x^\bullet)$. Meanwhile, we have $^{\bullet\bullet} x = \cup_{y \in {}^\bullet x} {}^\bullet y$ and $x^{\bullet\bullet} = \cup_{y \in x^\bullet} y^\bullet$.

EXAMPLE 2.4 For the Petri net in Fig. 2.1, it is clear that $^\bullet p_1 = \{t_4\}$, $p_1^\bullet = \{t_1\}$, $^\bullet p_2 = \{t_1\}$, $p_2^\bullet = \{t_4\}$, $^\bullet p_3 = \{t_1\}$, $p_3^\bullet = \{t_2, t_3\}$, $^\bullet p_4 = \{t_2\}$, $p_4^\bullet = \{t_3\}$, $^\bullet p_5 = \{t_3\}$, and $p_5^\bullet = \{t_4\}$. Moreover, $^\bullet \{p_1, p_2, p_3\} = \{t_1, t_4\}$ and $\{p_1, p_2, p_3\}^\bullet = \{t_1, t_2, t_3, t_4\}$.

Definition 2.5 A transition $t \in T$ is enabled at a marking M if $\forall p \in {}^\bullet t$, $M(p) \geq W(p, t)$. This fact is denoted as $M[t\rangle$. Firing it yields a new marking M' such that $\forall p \in P$, $M'(p) = M(p) - W(p, t) + W(t, p)$, denoted by $M[t\rangle M'$. M' is called an immediately reachable marking from M. Marking M'' is said to be reachable from M if there exists a sequence of transitions $\sigma = t_0 t_1 \cdots t_n$ and markings $M_1, M_2, \cdots$, and M_n such that $M[t_0\rangle M_1[t_1\rangle M_2 \cdots M_n[t_n\rangle M''$ holds. The set of markings reachable from M in N is called the reachability set of Petri net (N, M) and denoted as $R(N, M)$.

EXAMPLE 2.5 For the Petri net in Fig. 2.1, only t_1 is enabled at the initial marking, i.e., $M_0[t_1\rangle$. Firing t_1, a token is removed from p_1 and put two tokens in p_2 and one token in p_3. It yields a new marking $M_1 = 2p_2 + p_3$, denoted as $M_0[t_1\rangle M_1$. Similarly, we have $M_1[t_2\rangle M_2$, $M_2[t_2\rangle M_3$, and $M_3[t_2\rangle M_0$, where $M_2 = 2p_2 + p_4$ and $M_3 = 2p_2 + p_5$. Moreover, we have $M_0[t_1 t_2\rangle M_2$. The reachability set of the net is $R(N, M_0) = \{M_0, M_1, M_2, M_3\}$.

Definition 2.6 A Petri net (N, M_0) is safe if $\forall M \in R(N, M_0)$, $\forall p \in P$, $M(p) \le 1$ is true. It is bounded if $\exists k \in \mathbb{N}^+$, $\forall M \in R(N, M_0)$, $\forall p \in P$, $M(p) \le k$. It is said to be unbounded if it is not bounded. A net N is structurally bounded if it is bounded for any initial marking.

Definition 2.7 A net $N = (P, T, F, W)$ is pure (self-loop free) if $\forall x, y \in P \cup T$, $W(x, y) > 0$ implies $W(y, x) = 0$.

EXAMPLE 2.6 The Petri net in Fig. 2.1 is not safe since p_2 can hold two tokens when t_1 fires at the initial marking. The Petri net shown in Fig. 2.2 is safe. Both net models are bounded, structurally bounded, and pure.

Definition 2.8 The output and input incidence matrices of a net are defined as $[N^+](p, t) = W(t, p)$ and $[N^-](p, t) = W(p, t)$, respectively. The incidence matrix of a net is defined as $[N] = [N^+] - [N^-]$, where $[N](p, t) = W(t, p) - W(p, t)$. For a place p (transition t), its incidence vector, a row (column) in $[N]$, is denoted by $[N](p, \cdot)$ ($[N](\cdot, t)$).

A pure net $N = (P, T, F, W)$ and its incidence matrix $[N]$ have one-to-one relationship. However, for a non-pure net, its incidence matrix cannot represent self-loops. That is to say, two nets have the same incidence matrix but may have different net structures.

EXAMPLE 2.7 For the Petri net in Fig. 2.1, we have

$$[N^+] = \begin{bmatrix} 0 & 0 & 0 & 1 \\ 2 & 0 & 0 & 0 \\ 1 & 0 & 0 & 0 \\ 0 & 1 & 0 & 0 \\ 0 & 0 & 1 & 0 \end{bmatrix}, \quad [N^-] = \begin{bmatrix} 1 & 0 & 0 & 0 \\ 0 & 0 & 0 & 2 \\ 0 & 1 & 1 & 0 \\ 0 & 0 & 1 & 0 \\ 0 & 0 & 0 & 1 \end{bmatrix}, \quad [N] = \begin{bmatrix} -1 & 0 & 0 & 1 \\ 2 & 0 & 0 & -2 \\ 1 & -1 & -1 & 0 \\ 0 & 1 & -1 & 0 \\ 0 & 0 & 1 & -1 \end{bmatrix}. \quad (2.1)$$

Definition 2.9 Given a Petri net (N, M_0), $t \in T$ is live at M_0 if $\forall M \in R(N, M_0)$, $\exists M' \in R(N, M)$, $M'[t\rangle$. (N, M_0) is live if $\forall t \in T$, t is live at M_0. (N, M_0) is dead at M_0 if $\nexists t \in T$, $M_0[t\rangle$. (N, M_0) is deadlock-free (weakly live or live-lock) if $\forall M \in R(N, M_0)$, $\exists t \in T$, $M[t\rangle$.

Definition 2.10 Let $N = (P, T, F, W)$ be a net and σ be a finite sequence of transitions. The Parikh vector of σ is $\vec{\sigma} : T \to \mathbb{N}$ which maps t in T to the number of occurrences of t in σ. Denote $\vec{t_1} = (1\ 0\ \cdots\ 0)^T$, $\vec{t_2} = (0\ 1\ 0\ \cdots\ 0)^T$, and $\vec{t_k} = (0\ 0\ \cdots\ 0\ 1)^T$ assuming $k = |T|$.

It is trivial that for each transition t, we have $[N](\cdot,t) = [N]\vec{t}$. Note that $M[t\rangle M'$ leads to $M' = M + [N](\cdot,t)$. Consequently, if $M[t\rangle M'$, we have $M' = M + [N]\vec{t}$. For an arbitrary finite transition sequence σ such that $M[\sigma\rangle M'$, we have

$$M' = M + [N]\vec{\sigma} \qquad (2.2)$$

Eq. (2.2) is called the state equation of Petri net (N,M). It is just a necessary condition for the reachability of a marking M' from M if Eq. (2.2) has a non-negative integer solution. That is to say, any reachable marking satisfies the state equation but the converse is not true.

2.2.2 Structural Analysis

Structural properties are an important means to analyze and control Petri nets. Structural analysis is always applied via special structural objects of a Petri net, such as place invariants, siphons, and resource-transition circuits. Based on structural properties, the behavior of a Petri net can be analyzed simply. Also, the structural objects can always be easily found by mixed integer programs (MIPs) which are rather efficient. This section presents some classical structural objects of a Petri net and their properties.

Definition 2.11 A P-vector is a column vector $I : P \to \mathbb{Z}$ indexed by P and a T-vector is a column vector $J : T \to \mathbb{Z}$ indexed by T, where $\mathbb{Z}$ is the set of integers.

Definition 2.12 P-vector I is called a P-invariant (place invariant, PI for short) if $I \neq \mathbf{0}$ and $I^T[N] = \mathbf{0}^T$. T-vector J is called a T-invariant (transition invariant) if $J \neq \mathbf{0}$ and $[N]J = \mathbf{0}$.

Definition 2.13 P-invariant I is a P-semiflow if every element of I is non-negative. $\|I\| = \{p|I(p) \neq 0\}$ is called the support of I. $\|I\|^+ = \{p|I(p) > 0\}$ denotes the positive support of P-invariant I and $\|I\|^- = \{p|I(p) < 0\}$ denotes the negative support of I. I is called a minimal P-invariant if $\|I\|$ is not a superset of the support of any other one and its components are mutually prime.

Theorem 2.1 *Let (N,M_0) be a net with P-invariant I and M a reachable marking from M_0. Then*

$$I^T M = I^T M_0$$

Property 2.1 If I is a P-semiflow of a net, $^\bullet\|I\| = \|I\|^\bullet$.

EXAMPLE 2.8 For the Petri net in Fig. 2.3, there are two P-invariants $I_1 = (1\ 1\ 1\ 2\ 0)^T$ and $I_2 = (0\ 1\ 0\ 0\ 1)^T$, and one T-invariant $J_1 = (1\ 1\ 1\ 1)^T$. We have $\|I_1\| = \{p_1,p_2,p_3,p_4\}$ and $\|I_2\| = \{p_2,p_5\}$. Both I_1 and I_2 are minimal. According to Property 2.1, we also have $^\bullet\|I_1\| = \|I_1\|^\bullet = \{t_1,t_2,t_3,t_4\}$ and $^\bullet\|I_2\| = \|I_2\|^\bullet = \{t_1,t_2\}$.

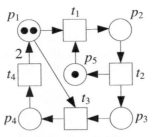

Figure 2.3 A Petri net example.

Definition 2.14 A nonempty set $S \subseteq P$ is a siphon if ${}^\bullet S \subseteq S^\bullet$. $S \subseteq P$ is a trap if $S^\bullet \subseteq {}^\bullet S$. A siphon (trap) is minimal if there is no siphon (trap) contained in it as a proper subset. A minimal siphon S is said to be strict if ${}^\bullet S \subsetneq S^\bullet$.

Property 2.2 Let S_1 and S_2 be two siphons (traps). Then, $S_1 \cup S_2$ is a siphon (trap).

EXAMPLE 2.9 In the Petri net in Fig. 2.3, there is a siphon $S = \{p_1, p_4\}$ since ${}^\bullet S \subset S^\bullet$, where ${}^\bullet S = \{t_3, t_4\}$ and $S^\bullet = \{t_1, t_3, t_4\}$. In fact, S is both strict and minimal since ${}^\bullet S \subset S^\bullet$ and there is no other siphon contained in S as a proper subset.

Corollary 2.1 *If I is a P-semiflow, then $\|I\|$ is both a siphon and trap.*

EXAMPLE 2.10 In the Petri net in Fig. 2.3, $S = \{p_2, p_5\}$ is both a siphon and trap since ${}^\bullet S = S^\bullet = \{t_1, t_2\}$. S is also the support of a P-semiflow.

The sum of tokens in a set S is denoted as $M(S) = \sum_{p \in S} M(p)$. S is marked at M if $M(S) > 0$ and unmarked if $M(S) = 0$.

Property 2.3 Let $M \in R(N, M_0)$ be a marking of net (N, M_0) and S a trap. If $M(S) > 0$, then $\forall M' \in R(N, M)$, $M'(S) > 0$.

Property 2.4 Let $M \in R(N, M_0)$ be a marking of net (N, M_0) and S a siphon. If $M(S) = 0$, then $\forall M' \in R(N, M)$, $M'(S) = 0$.

Property 2.3 indicates that once a trap is marked at M, it is always marked at any reachable marking from M. Property 2.4 shows that once a siphon is empty at M, it remains empty at any reachable marking from M.

Theorem 2.2 *Let (N, M_0) be an ordinary net and Π the set of its siphons. The net is deadlock-free if $\forall S \in \Pi$, $\forall M \in R(N, M_0)$, $M(S) > 0$.*

Definition 2.15 A siphon S is said to be controlled in a net system (N, M_0) if $\forall M \in R(N, M_0)$, $M(S) > 0$.

Definition 2.16 Let $N = (P, T, F, W)$ be a Petri net with $P_X \subseteq P$ and $T_X \subseteq T$. $N_X = (P_X, T_X, F_X, W_X)$ is called a subnet generated by $P_X \cup T_X$ if $F_X = F \cap [(P_X \times T_X) \cup (T_X \times P_X)]$ and $\forall f \in F_X$, $W_X(f) = W(f)$.

EXAMPLE 2.11 Fig. 2.4 shows two subnets of the Petri net shown in Fig. 2.3, where N_1 is generated by $P_{X_1} = \{p_2, p_5\}$ and $T_{X_1} = \{t_1, t_2\}$, and N_2 is generated by $P_{X_2} = \{p_1, p_2, p_5\}$ and $T_{X_2} = \{t_1, t_4\}$.

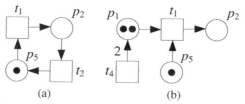

Figure 2.4 (a) a subnet N_1 generated by $\{p_2, p_5\} \cup \{t_1, t_2\}$ and (b) a subnet N_2 generated by $\{p_1, p_2, p_5\} \cup \{t_1, t_4\}$.

2.2.3 Reachability Graph

The reachability graph of a Petri net can reflect its complete behavior, i.e., it exactly presents the evolution of the net system. Thus, it is important for the control of a Petri net, especially its optimal control. This section shows the analysis of a reachability graph from the viewpoint of deadlock control.

Let $G(N, M_0)$ be the reachability graph of a bounded Petri net. For deadlock control purposes, markings in a reachability graph can be classified into four categories—deadlock, bad, dangerous and good ones. A deadlock one indicates a dead situation in a system by which no successor is followed. A bad one has successors but cannot reach the initial marking. A good marking is the one which can reach the initial one and whose successors can also reach it. A dangerous marking can reach the initial one and at least one of its successors cannot reach it. For the optimal control purposes, both good and dangerous markings should be kept in the controlled system, which therefore are legal markings whose set is denoted by $\mathcal{M}_L$. Fig. 2.5 shows the reachability graph of a Petri net, where M_{13} and M_{14} are deadlock markings, M_4, M_8, and M_9 are bad, M_1, M_2, M_3, M_5, M_6, and M_{11} are dangerous, and the others are good. The set of good and dangerous markings in $R(N, M_0)$, namely $\mathcal{M}_L$, should constitute the maximum legal behavior if a supervisor is optimally designed.

For a Petri net system (N, M_0), the set of its legal markings is defined as

$$\mathcal{M}_L = \{M | M \in R(N, M_0) \wedge M_0 \in R(N, M)\} \tag{2.3}$$

The set $\mathcal{M}_L$ is the maximal set of reachable markings such that it is possible to reach initial marking M_0 from any legal marking without leaving $\mathcal{M}_L$. In (Uzam and Zhou, 2006, 2007), $G(N, M_0)$ is split into a deadlock-zone (DZ) and a live-zone (LZ), where the DZ contains deadlock and bad markings and the LZ contains all the legal markings. A first-met bad marking (FBM) is defined as the one within DZ, representing the very first entry from the LZ to the DZ. A mathematical form of the set of first-met bad markings is defined as

$$\mathcal{M}_{\mathrm{FBM}} = \{M | M \text{ in } DZ, \exists M' \text{ in } LZ, t \in T, \text{s.t. } M'[t\rangle M\} \tag{2.4}$$

$M_0 = 3p_1 + 3p_8 + p_9 + p_{10} + p_{11}$

$M_1 = 2p_1 + p_2 + 3p_8 + p_{10} + p_{11}$

$M_2 = 3p_1 + p_5 + 2p_8 + p_9 + p_{10}$

$M_3 = 2p_1 + p_3 + 3p_8 + p_9 + p_{11}$

$M_4 = 2p_1 + p_2 + p_5 + 2p_8 + p_{10}$

$M_5 = 3p_1 + p_6 + 2p_8 + p_9 + p_{11}$

$M_6 = p_1 + p_2 + p_3 + 3p_8 + p_{11}$

$M_7 = 2p_1 + p_4 + 3p_8 + p_9 + p_{10}$

$M_8 = 2p_1 + p_3 + p_5 + 2p_8 + p_9$

$M_9 = 2p_1 + p_2 + p_6 + 2p_8 + p_{11}$

$M_{10} = 3p_1 + p_7 + 2p_8 + p_{10} + p_{11}$

$M_{11} = 3p_1 + p_5 + p_6 + p_8 + p_9$

$M_{12} = p_1 + p_2 + p_4 + 3p_8 + p_{10}$

$M_{13} = p_1 + p_2 + p_3 + p_5 + 2p_8$

$M_{14} = 2p_1 + p_2 + p_5 + p_6 + p_8$

$M_{15} = 3p_1 + p_5 + p_7 + p_8 + p_{10}$

$M_{16} = p_1 + p_3 + p_4 + 3p_8 + p_9$

$M_{17} = 3p_1 + p_6 + p_7 + p_8 + p_{11}$

$M_{18} = p_2 + p_3 + p_4 + 3p_8$

$M_{19} = 3p_1 + p_5 + p_6 + p_7$

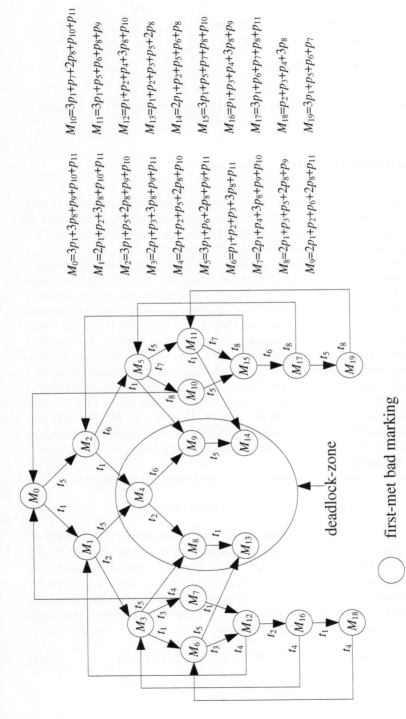

deadlock-zone

first-met bad marking

Figure 2.5 The reachability graph of a Petri net.

We can see that from an FBM, it is not possible to reach the initial marking and it may lead to a deadlock one or a group of bad ones. In Fig. 2.5, M_4, M_8, M_9, M_{13}, and M_{14} are FBMs. For an FBM, in (Uzam and Zhou, 2006), the authors consider the markings of operation places only and prevent the marking of the subset of the operation places from being reached. The marking of the subset of the operation places is characterized as a PI that can be implemented by a control place.

2.3 BINARY DECISION DIAGRAMS

Binary Decision Diagrams (BDDs) are considered as a powerful tool to represent a large number of encoded data with a compressed shared data structure and can implement the efficient operations on the data. In the deadlock analysis of a Petri net, BDDs can be used to compute all minimal siphons and the reachability graph that are two key concepts for the deadlock control purposes. This section briefly sketches the fundamental concepts of BDDs. More details can be found in (Andersen, 1997; Brant, 1992; Pastor *et al.*, 1994, 2001).

2.3.1 Boolean Algebra

Definition 2.17 A set is a collection of objects called elements. The cardinality of a set A, denoted by $|A|$, is the number of elements in it. The power set of A is the set of all subsets of the elements in A, which is represented by 2^A.

Definition 2.18 Let A and B be two sets. A binary relation $\mathcal{R}$ between A and B is a subset of the cartesian product $A \times B$. $\forall x \in A$ and $\forall y \in B$, $x\mathcal{R}y$ denotes that x and y are in relation $\mathcal{R}$. A function f from A to B, denoted as $f : A \rightarrow B$, is a relation that associates exactly one element of B with each element of A. In this case, A and B are called the domain and codomain of the function, respectively. $\forall x \in A$, $f(x) \in B$ is called the image of x.

Definition 2.19 A Boolean algebra is a five-tuple $(B, +, \cdot, 0, 1)$, where B is a set called the carrier, $+$ and $\cdot$ are binary operations on B, and 0 and 1 are elements of B. The elements in B satisfy the commutative, distributive, identity, and complement laws, as presented below. $\forall a, b, c \in B$:

1. Commutative laws: $a + b = b + a$ and $a \cdot b = b \cdot a$.
2. Distributive laws: $a + (b \cdot c) = (a + b) \cdot (a + c)$ and $a \cdot (b + c) = a \cdot b + a \cdot c$.
3. Identity laws: $a + 0 = a$ and $a \cdot 1 = a$.
4. Complement laws: $\forall a \in B$, $\exists \bar{a} \in B$, called the complementary such that $a + \bar{a} = 1$ and $a \cdot \bar{a} = 0$.

Definition 2.20 The algebra of the subsets of a set S, denoted by $(2^S, \bigcup, \bigcap, \emptyset, S)$, is a Boolean algebra, where 2^S is the set of all the subsets of S, $\bigcup$ and $\bigcap$ are the union and intersection operations, respectively, and $\emptyset$ is the empty set.

Definition 2.21 Let $n \geq 0$ be an integer. An n-variable Boolean function is a function $f_n : B^n \to B$. Let $F_n(B)$ be the power set of B^n (the characteristic functions of subsets of B^n). The system $(F_n(B), +, \cdot, \mathbf{0}, \mathbf{1})$ is the Boolean algebra of Boolean functions, in which "+" and "·" represent disjunction and conjunction of n-variable Boolean functions and $\mathbf{0}$ and $\mathbf{1}$ represent the *"zero"* and *"one"* functions $(f(x_1,\ldots,x_n) = 0$ and $f(x_1, \ldots, x_n) = 1)$, respectively.

Let $V \subseteq B^n$ be a set of elements in the Boolean algebra of n-variable Boolean functions. The characteristic function X_V of the set V is an n-variable Boolean function that evaluates to 1 for those elements of B^n that are in V, i.e., $v \in V \Leftrightarrow X_V(v) = 1, \forall v \in F_n(B)$.

Theorem 2.3 (Stone's Representation Theorem) *Every finite Boolean algebra is isomorphic to the Boolean algebra of the subsets of some finite set S.*

Theorem 2.3 indicates that bounded Petri nets can be modeled with the Boolean algebra of Boolean functions. Thus, the symbolic computation and analysis of Petri nets can be accordingly achieved by using BDDs, which is shown in Chapter 3.

Definition 2.22 Let $x_1, x_2, \ldots$, and x_n be the n variables of n-variable Boolean functions. Each element of B^n is called a vertex. A literal is either a variable x_i or its complement $\bar{x}_i$. A cube c is a set of literals such that if $x_i \in c$, then $\bar{x}_i \notin c$ and if $\bar{x}_i \in c$, then $x_i \notin c$. In this case, the set of all cubes with n literals is in one-to-one correspondence with the vertices of B^n.

EXAMPLE 2.12 Let x_1, x_2, x_3, x_4, and x_5 be the variables of Boolean function f_n : $B^5 \to B$. $(1\ 0\ 1\ 1\ 0)^T \in B^5$ is a vertex. A cube $c = x_1 \bar{x}_2 x_3 x_4 \bar{x}_5$ is the one-to-one correspondence with $(1\ 0\ 1\ 1\ 0)^T$.

Boolean functions $f_{x_i} = f(x_1, \ldots, x_{i-1}, 1, x_{i+1}, \ldots, x_n)$ and $f_{\bar{x}_i} = f(x_1, \ldots, x_{i-1}, 0, x_{i+1}, \ldots, x_n)$ are called the positive and negative cofactors of f with respect to x_i, respectively. If $f : B^n \to B$ is an n-variable Boolean function, then we have

$$\forall x_i, 1 \leq i \leq n, f(x_1, \ldots, x_n) = x_i \cdot f_{x_i} + \bar{x}_i \cdot f_{\bar{x}_i}$$

EXAMPLE 2.13 Let $f(x_1, x_2, x_3) = x_1 x_2 + \bar{x}_1 x_2 x_3 + \bar{x}_2 x_3$. Then, we have $f_{x_1} = x_2 + \bar{x}_2 x_3$, $f_{x_2} = x_1 + \bar{x}_1 x_3$, $f_{x_3} = x_1 x_2 + \bar{x}_1 x_2 + \bar{x}_2$, $f_{\bar{x}_1} = x_2 x_3 + \bar{x}_2 x_3$, $f_{\bar{x}_2} = x_3$, and $f_{\bar{x}_3} = x_1 x_2$. It is easy to verify that $x_1 \cdot f_{x_1} + \bar{x}_1 \cdot f_{\bar{x}_1} = x_1 \cdot (x_2 + \bar{x}_2 x_3) + \bar{x}_1 \cdot (x_2 x_3 + \bar{x}_2 x_3) = x_1 x_2 + x_1 \bar{x}_2 x_3 + \bar{x}_1 x_2 x_3 + \bar{x}_1 \bar{x}_2 x_3 = x_1 x_2 + \bar{x}_1 x_2 x_3 + (x_1 \bar{x}_2 x_3 + \bar{x}_1 \bar{x}_2 x_3) = x_1 x_2 + \bar{x}_1 x_2 x_3 + \bar{x}_2 x_3$, i.e., $f(x_1, x_2, x_3) = x_1 \cdot f_{x_1} + \bar{x}_1 \cdot f_{\bar{x}_1}$. Similarly, we can verify that $f(x_1, x_2, x_3) = x_2 \cdot f_{x_2} + \bar{x}_2 \cdot f_{\bar{x}_2}$ and $f(x_1, x_2, x_3) = x_3 \cdot f_{x_3} + \bar{x}_3 \cdot f_{\bar{x}_3}$.

Definition 2.23 Let f be a Boolean function with respect to x_i. The existential and universal abstractions of f are defined as $\exists_{x_i} f = f_{x_i} + f_{\bar{x}_i}$ and $\forall_{x_i} f = f_{x_i} \cdot f_{\bar{x}_i}$, respectively.

EXAMPLE 2.14 Let $f(x_1, x_2, x_3) = x_1 x_2 + \bar{x}_1 x_2 x_3 + \bar{x}_2 x_3$. Then, we have $\exists_{x_1} f(x_1, x_2, x_3) = x_2 + x_2 x_3 + \bar{x}_2 x_3$ and $\forall_{x_1} f(x_1, x_2, x_3) = x_2 x_3 + \bar{x}_2 x_3 = x_3$.

2.3.2 Binary Decision Diagrams

This section reviews some basic definitions for BDDs. More details can be found in (Andersen, 1997).

Definition 2.24 (Andersen, 1997) A Binary Decision Diagram (BDD) is a rooted, directed acyclic graph with two sink nodes, labeled by 0 and 1, representing the Boolean functions **0** and **1**, respectively. Each nonsink node is labeled with a Boolean variable v and has two out-edges labeled by 0 and 1. Each nonsink node represents the Boolean function corresponding to its 0-edge if $v = 0$ or the Boolean function corresponding to its 1-edge if $v = 1$. (In graphical representation, they are shown as dotted and solid lines, respectively.)

Definition 2.25 (Andersen, 1997) A BDD is ordered (Ordered Binary Decision Diagram, OBDD for short) if on all paths through the graph the variables respect a given linear order. An OBDD is reduced (Reduced Ordered Binary Decision Diagram, ROBDD for short) if no two distinct nodes represent the same logic function and no variable node has identical 1-edge and 0-edge.

ROBDDs have very important properties. They provide compact representations of Boolean expressions, and there are efficient algorithms for performing all kinds of logical operations on ROBDDs. They are all based on the crucial fact that for any function $f : B^n \to B$ there is exactly one ROBDD representing it. In the following, we use "BDD" to imply "ROBDD". For example, given a logic function:

$$f = abc + \bar{a}\bar{b} + \bar{b}c$$

its OBDD and ROBDD are shown in Fig. 2.6 with an order of $a < b < c$.

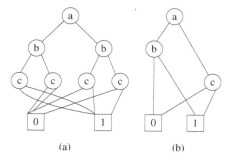

(a)   (b)

Figure 2.6 BDD examples: (a) OBDD and (b) ROBDD with the order $a < b < c$.

The order of variables in a ROBDD greatly influences its structural complexity and the efficiency of its computation. For example, the ROBDD shown in Fig. 2.6 can also be presented with the orders $c < b < a$ and $c < a < b$, as dipicted in Fig. 2.7. It can be seen that different orders may lead to different BDD structures. For this example, the BDD with the order $a < b < c$ has the simplest structure.

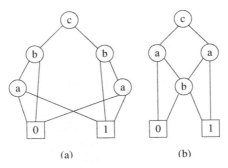

(a)   (b)

Figure 2.7 (a) ROBDD with the order $c < b < a$ and (b) ROBDD with the order $c < a < b$.

BDDs are very powerful to represent a large set of encoded data with a compact data structure. Thus, they are widely used to overcome computational complexity problems in applications. Researchers and engineers develop a number of software packages to implement BDD operations by programming in computers. Representative tools are BuDDy (Lind-Nielsen, 2002), CUDD (Somenzi, 2005), Spot, etc. With the help of these software packages, one can easily develop their own programs to implement BDD operations.

2.4 BIBLIOGRAPHICAL REMARKS

The basic definitions of Petri nets can be found in books (Desel and Esparza, 1995; Li and Zhou, 2009; Peterson, 1981; Reisig, 1985; Zhou and Venkatesh, 1998) and survey papers (Murata, 1989; Peterson, 1977). Efficient algorithms to compute siphons can be found in (Chao, 2006a,b; Cordone *et al.*, 2005; Ezpeleta *et al.*, 1993; Wang *et al.*, 2009). The introduction to BDDs refers to (Andersen, 1997; Brant, 1992), and their applications to the analysis of Petri nets can be found in (Pastor *et al.*, 1994, 2001). A good tool to analyze the structure and reachability graph of a Petri net model refers to (Starke, 2003).

References

Abdallah, I. B. and H. A. ElMaraghy. 1998. Deadlock prevention and avoidance in FMS: A Petri net based approach. International Journal of Advanced Manufacturing Technology. 14(10): 704–715.

Andersen, H. R. 1997. An introduction to binary decision diagrams. Lecture Notes for 49285 Advanced Algorithms E97, Department of Information Technology, Technical University of Denmark.

Brant, R. 1992. Symbolic Boolean manipulation with ordered binary decision diagrams. ACM Computing Surveys. 24(3): 293–318.

Chao, D. Y. 2006a. Computation of elementary siphons in Petri nets for deadlock control. Computer Journal. 49(4): 470–479.

Chao, D. Y. 2006b. Searching strict minimal siphons for SNC-based resource allocation systems. Journal of Information Science and Engineering. 23(3): 853–867.

Chen, Y. F. and Z. W. Li. 2011. Design of a maximally permissive liveness-enforcing supervisor with a compressed supervisory structure for flexible manufacturing systems. Automatica. 47(5): 1028–1034.

Chen, Y. F., Z. W. Li, M. Khalgui, and O. Mosbahi. 2011. Design of a maximally permissive liveness-enforcing Petri net supervisor for flexible manufacturing systems. IEEE Transactions on Automation Science and Engineering. 8(2): 374–393.

Coffman, E. G., M. J. Elphick, and A. Shoshani. 1971. System deadlocks. ACM Computing Surveys. 3(2): 67–78.

Cordone, R., L. Ferrarini, and L. Piroddi. 2005. Enumeration algorithms for minimal siphons in Petri nets based on place constraints. IEEE Transactions on System, Man and Cybernetics, Part A. 35(6): 844–854.

Desel, J. and J. Esparza. 1995. Free Choice Petri Nets. Cambridge University Press, London, UK.

Ezpeleta, J., J. M. Couvreur, and M. Silva. 1993. A new technique for finding a generating family of siphons, traps, and st-components: Application to colored Petri nets. Lecture Notes in Computer Science. 674: 126–147.

Ezpeleta, J., J. M. Colom, and J. Martinez. 1995. A Petri net based deadlock prevention policy for flexible manufacturing systems. IEEE Transactions on Robotics and Automation. 11(2): 173–184.

Somenzi, F. 2005. CUDD: CU Decision Diagram Package Release 2.4.2. University of Colorado at Boulder.

Fanti, M. P., B. Maione, and B. Turchiano. 2000. Comparing digraph and Petri net approaches to deadlock avoidance in FMS. IEEE Transactions on Systems, Man and Cybernetics, Part B. 30(5): 783–798.

Fanti, M. P. and M. C. Zhou. 2004. Deadlock control methods in automated manufacturing systems. IEEE Transactions on Systems, Man, and Cybernetics, Part A. 34(1): 5–22.

Fanti, M. P. and M. C. Zhou. Deadlock control methods in automated manufacturing systems. pp. 1–22. In M. C. Zhou and M. P. Fanti. [eds.]. 2005. Deadlock Resolution in Computer-Integrated Systems. New York: Marcel-Dekker Co.

Ghaffari, A., N. Rezg, and X. L. Xie. 2003. Design of a live and maximally permissive Petri net controller using the theory of regions. IEEE Transactions on Robotics and Automation. 19(1): 137–142.

Hu, H. S., M. C. Zhou, and Z. W. Li. 2010. Low-cost high-performance supervision in ratio-enforced automated manufacturing systems using timed Petri nets. IEEE Transactions on Automation Science and Engineering. 7(4): 933–944.

Huang, Y. S., M. D. Jeng, X. L. Xie, and S. L. Chung. 2001. Deadlock prevention based on Petri nets and siphons. International Journal of Production Research. 39(2): 283–305.

Huang, Y. S., M. D. Jeng, X. L.Xie, and D. H. Chung. 2006. Siphon-based deadlock prevention for flexible manufacturing systems. IEEE Transactions on Systems, Man, and Cybernetics, Part A. 36(6): 1248–1256.

Jeng, M. D. and X. L. Xie. Deadlock detection and prevention of automated manufacturing systems using Petri nets and siphons. pp. 233–281. In M. C. Zhou and M. P. Fanti. [eds.]. 2005. Deadlock Resolution in Computer-Integrated Systems. New York: Marcel-Dekker Inc.

Li, Z. W., H. S. Hu, and A. R. Wang. 2007. Design of liveness-enforcing supervisors for flexible manufacturing systems using Petri nets. IEEE Transactions on Systems, Man, and Cybernetics, Part C. 37(4): 517–526.

Li, Z. W. and M. C. Zhou. 2004. Elementary siphons of Petri nets and their application to deadlock prevention in flexible manufacturing systems. IEEE Transactions on Systems, Man, and Cybernetics, Part A. 34(1): 38–51.

Li, Z. W. and M. C. Zhou. 2009. Deadlock Resolution in Automated Manufacturing Systems: A Novel Petri Net Approach. Springer, London, UK.

Lind-Nielsen, J. 2002. BuDDy: Binary Decision Diagram Package Release 2.2. IT-University of Copenhagen (ITU).

Miner, A. S. and G. Ciardo. 1999. Efficient reachability set generation and storage using decision diagrams. Lecture Notes in Computer Science. 1639: 6–25.

Murata, T. 1989. Petri nets: Properties, analysis and application. Proceedings of the IEEE. 77(4): 541–580.

Pastor, E., J. Cortadella, and O. Roig. 2001. Symbolic analysis of bounded Petri nets. IEEE Transactions on Computers. 50(5): 432–448.

Pastor, E., O. Roig, J. Cortadella, and R. M. Badia. 1994. Petri net analysis using Boolean manipulation. Lecture Notes in Computer Science. 815: 416–435.

Peterson, J. L. 1977. Petri nets. Computing Surveys. 9(3): 223–252.

Peterson, J. L. 1981. Petri Net Theory and the Modeling of Systems. Englewood Cliffs: Prentice-Hall.

Piroddi, L., R. Cordone, and I. Fumagalli. 2008. Selective siphon control for deadlock prevention in Petri nets. IEEE Transactions on Systems, Man, and Cybernetics, Part A. 38(6): 1337–1348.

Piroddi, L., R. Cordone, and I. Fumagalli. 2009. Combined siphon and marking generation for deadlock prevention in Petri nets. IEEE Transactions on Systems, Man and Cybernetics, Part A. 39(3): 650–661.

Ramadge, P. and W. M. Wonham. 1987. Supervisory control of a class of discrete event processes. SIAM Journal on Control and Optimization. 25(1): 206–230.

Ramadge, P. and W. M. Wonham. 1989. The control of discrete event systems. Proceedings of the IEEE. 77(1): 81–89.

Reisig, W. 1985. Petri Nets: An Introduction. Springer-Verlag, Newyork USA.

Reveliotis, S. A. 2007. Implicit siphon control and its role in the liveness-enforcing supervision of sequential resource allocation systems. IEEE Transactions on Systems, Man, and Cybernetics, Part A. 37(3): 319–328.

Starke, P. H. 2003. INA: Integrated Net Analyzer. http://www2.informatik.huberlin.de/starke/ina.html.

Uzam, M. and M. C. Zhou. 2006. An improved iterative synthesis method for liveness enforcing supervisors of flexible manufacturing systems. International Journal of Production Research. 44(10): 1987–2030.

Uzam, M. and M. C. Zhou. 2007. An iterative synthesis approach to Petri net based deadlock prevention policy for flexible manufacturing systems. IEEE Transactions on Systems, Man and Cybernetics, Part A. 37(3): 362–371.

Wang, A. R., Z. W. Li, J. Y. Jia, and M. C. Zhou. 2009. An effective algorithm to find elementary siphons in a class of Petri nets. IEEE Transactions on Systems, Man, and Cybernetics, Part A. 39(4): 912–923.

Wikipedia. Petri Net. http://en.wikipedia.org/wiki/Petri_net.

Wu, N. Q. 1999. Necessary and sufficient conditions for deadlock-free operation in flexible manufacturing systems using a colored Petri net model. IEEE Transactions on Systems, Man, and Cybernetics, Part C. 29(2): 192–204.

Wu, N. Q. and M. C. Zhou. 2010. System Modeling and Control with Resource-Oriented Petri Nets. CRC Press, New York.

Zhou, M. C. and F. DiCesare. 1991. Parallel and sequential mutual exclusions for Petri net modeling of manufacturing systems with shared resources. IEEE Transactions on Robotics and Automation. 7(4): 515–527.

Zhou, M. C., F. DiCesare, and A. A. Desrochers. 1992. A hybrid methodology for synthesis of Petri nets for manufacturing systems. IEEE Transactions on Robotics and Automation. 8(3): 350–361.

Zhou, M. C. and F. DiCesare. 1993. Petri Net Synthesis for Discrete Event Control of Manufacturing Systems. Kluwer Academic Publishers, London, UK.

Zhou, M. C. and K. Venkatesh. 1998. Modeling, Simulation, and Control of Flexible Manufacturing Systems: A Petri Net Approach. World Scientific, Singapore.

Chapter 3

Symbolic Computation and Analysis of Petri Nets

ABSTRACT

Reachability graph analysis of Petri nets is a key technique for the deadlock control in AMSs. However, the number of reachable markings increases exponentially with respect to the size of a Petri net. Thus, a complete enumeration of markings of a Petri net always suffers from the state explosion problem. It often leads to two problems in the process of computation on a computer: a very long computation time and the requirement of too much memory for saving the data. BDDs have the capability of representing a large set of encoded data by using a compact data structure and implementing the efficient manipulation of the set. This chapter presents the symbolic computation of reachable markings and an algorithm to find legal markings and FBMs. The BDD-based Petri net analysis technique makes more efficient the methods. We also present a symbolic approach to compute all minimal siphons of a Petri net in this chapter.

3.1 INTRODUCTION

Reachability graph analysis and structural analysis are two important Petri net analysis techniques for deadlock control. Based on the reachability graph of a Petri net model, an optimal supervisor can always be obtained since it carries the complete behavior information of a net model (Uzam, 2002; Uzam and Zhou, 2006). However, the number of reachable markings increases exponentially with respect to its size and the initial marking. A complete state enumeration always suffers from the state explosion problem. The computation of a reachability graph is known to be very time-consuming or even impossible. Based on the structural analysis, a deadlock control policy is generally carried out in terms of a special structure of a net (Wu, 1999; Wu and Zhou, 2001; Xing *et al.*, 1996; Yamalidou *et al.*, 1996). Siphons, particularly minimal siphons, are one of the special structures. As a structural object of Petri nets, siphons are closely related to the deadlocks. Thus, they are widely used in the development of deadlock control policies (Ezpeleta *et al.*, 2004; Ghaffari *et al.*, 2003; Huang *et al.*, 2001; Jeng and Xie, 2005; Kumaran *et al.*, 1994; Li *et al.*, 2007, 2008; Li and Zhou, 2004, 2006; Piroddi *et al.*, 2008;

Tricas *et al.*, 2000; Zhou and DiCesare, 1991). However, the number of siphons in a Petri net increases exponentially with respect to its size. Thus, the calculation of all siphons is computationally expensive. Even all the siphons are found, identifying minimal siphons from a large number of siphons is still time-consuming. This fact motivates us to find more efficient approaches to compute minimal siphons.

For bounded Petri nets, Pastor *et al.* provide a very efficient symbolic approach to compute the set of reachable markings (Pastor *et al.*, 1994, 2001) by using BDDs that have a compact and normative data structure. BDDs have the capability of representing large sets of encoded data with a small data structure and enable the efficient manipulation of these sets (Andersen, 1997; Brant, 1992). Thus, they are a very powerful tool to model and analyze Petri nets. This chapter presents the symbolic computation of reachable markings and an algorithm to find legal markings and FBMs by using BDDs. This makes more efficient the presented methods. Finally, we also show a symbolic approach to compute all minimal siphons of a Petri net.

3.2 SYMBOLIC MODELING OF BOUNDED PETRI NETS

This section first reports how a safe net can be modeled by using Boolean algebras. Then, the model is extended to bounded nets. More details about the symbolic modeling of Petri nets can be found in (Pastor *et al.*, 1994, 2001).

Generally, a marking in a safe Petri net can be represented by a set of places $M = \{p_1, \ldots, p_k\}$, where $p \in M$ denotes the fact that there is a token in p. Let $\mathcal{M}_P = 2^P$ be the set of all the subsets of the places representing markings in a safe net. In this case, the system $(2^{\mathcal{M}_P}, \cup, \cap, \emptyset, \mathcal{M}_P)$ is the Boolean algebra of the sets of markings. According to Theorem 2.3, the system is isomorphic to the Boolean algebra of n-variable Boolean functions, where n is the number of the places in P. Then, for any marking in $\mathcal{M}_P$ and a vector in B^n, there is a one-to-one correspondence. That is to say, any marking $M \in \mathcal{M}_P$ in a safe net can be represented by a vertex of B^n. A function $\varepsilon \colon \mathcal{M}_P \to B^n$ is used to encode every marking $M \in \mathcal{M}_P$ into a vertex $(p_1 \ldots p_n) \in B^n$ as follows:

$$\forall i \in 1, \ldots, n, \quad p_i = \begin{cases} 1 & \text{if } p_i \in M, \\ 0 & \text{if } p_i \notin M. \end{cases}$$

where p_i is used to denote either a place in M or its corresponding Boolean variable and M to denote either a reachable marking or the corresponding set of places that hold tokens at the marking. Then, each set of markings $\mathcal{M} \in 2^{\mathcal{M}_P}$ can be represented by a corresponding image $V \in F_n(B)$ according to ε, which is defined as follows:

$$V = \{v \in B^n | \exists M \in \mathcal{M} : v = \varepsilon(M)\}$$

Then the set $\mathcal{M}$ can be represented by those vertices for which function $\mathcal{X}_{\mathcal{M}} \colon B^n \to B$ evaluates to 1, where $\mathcal{X}_{\mathcal{M}}$ is called the characteristic function of $\mathcal{M}$. Thus, $\mathcal{X}_{\mathcal{M}} = \mathcal{X}_V$. Then, all set manipulations can be applied as Boolean operations directly to the characteristic functions. Let $\mathcal{M}_1$ and $\mathcal{M}_2$ be two sets of markings in $\mathcal{M}_P$. We have union $\mathcal{X}_{\mathcal{M}_1 \cup \mathcal{M}_2} = \mathcal{X}_{\mathcal{M}_1} + \mathcal{X}_{\mathcal{M}_2}$, intersection $\mathcal{X}_{\mathcal{M}_1 \cap \mathcal{M}_2} = \mathcal{X}_{\mathcal{M}_1} \cdot \mathcal{X}_{\mathcal{M}_2}$, and complement $\mathcal{X}_{\overline{\mathcal{M}_1}} = \overline{\mathcal{X}_{\mathcal{M}_1}}$.

EXAMPLE 3.1 Let $\mathcal{M}_1 = \{M_1, M_2, M_3\}$ and $\mathcal{M}_2 = \{M_3, M_4\}$ be two sets of markings in a safe Petri net with six places, where $M_1 = \{p_2, p_3\}$, $M_2 = \{p_2, p_6\}$, $M_3 = \{p_4, p_6\}$, and $M_4 = \{p_4, p_5\}$. Each marking has a correspondence in B^6, i.e., $v_1 = \varepsilon(M_1) = (0\ 1\ 1\ 0\ 0\ 0)^T$, $v_2 = \varepsilon(M_2) = (0\ 1\ 0\ 0\ 0\ 1)^T$, $v_3 = \varepsilon(M_3) = (0\ 0\ 0\ 1\ 0\ 1)^T$, and $v_4 = \varepsilon(M_4) = (0\ 0\ 0\ 1\ 1\ 0)^T$. Let $p_1, p_2, \ldots,$ and p_6 denote either the six places in the net or their corresponding variables. Then, the four markings can also be represented by cubes $c_1 = \bar{p}_1 p_2 p_3 \bar{p}_4 \bar{p}_5 \bar{p}_6$, $c_2 = \bar{p}_1 p_2 \bar{p}_3 \bar{p}_4 \bar{p}_5 p_6$, $c_3 = \bar{p}_1 \bar{p}_2 \bar{p}_3 p_4 \bar{p}_5 p_6$, and $c_4 = \bar{p}_1 \bar{p}_2 \bar{p}_3 p_4 p_5 \bar{p}_6$, respectively. The characteristic functions of sets $\mathcal{M}_1$ and $\mathcal{M}_2$ are $\mathcal{X}_{\mathcal{M}_1} = \bar{p}_1 p_2 \bar{p}_4 \bar{p}_5 (p_3 \oplus p_6) + \bar{p}_1 \bar{p}_3 \bar{p}_5 p_6 (p_2 \oplus p_4)$ and $\mathcal{X}_{\mathcal{M}_2} = \bar{p}_1 \bar{p}_2 \bar{p}_3 p_4 (p_5 \oplus p_6)$. For set manipulations, we have $\mathcal{X}_{\mathcal{M}_1 \cup \mathcal{M}_2} = \mathcal{X}_{\mathcal{M}_1} + \mathcal{X}_{\mathcal{M}_2} = \bar{p}_1 p_2 \bar{p}_4 \bar{p}_5 (p_3 \oplus p_6) + \bar{p}_1 \bar{p}_3 \bar{p}_5 p_6 (p_2 \oplus p_4) + \bar{p}_1 \bar{p}_2 \bar{p}_3 p_4 (p_5 \oplus p_6)$ and $\mathcal{X}_{\mathcal{M}_1 \cap \mathcal{M}_2} = \mathcal{X}_{\mathcal{M}_1} \cdot \mathcal{X}_{\mathcal{M}_2} = \bar{p}_1 \bar{p}_2 \bar{p}_3 p_4 \bar{p}_5 p_6$, which coincides with the fact that $\mathcal{M}_1 \cup \mathcal{M}_2 = \{M_1, M_2, M_3, M_4\}$ and $\mathcal{M}_1 \cap \mathcal{M}_2 = \{M_3\}$.   □

Binary relations between sets of markings can also be represented by characteristic functions. Let $\mathcal{M}$ and $\mathcal{M}'$ be two sets. A binary relation $\mathcal{R} \subseteq \mathcal{M} \times \mathcal{M}'$ can be represented by using two sets of Boolean variables $p_1, \ldots, p_n$ and $q_1, \ldots, q_n$ to encode the elements of the two sets $\mathcal{M}$ and $\mathcal{M}'$, respectively. Then, the characteristic function of $\mathcal{R}$ can be defined as:

$$\forall (p_1, \ldots, p_n), (q_1, \ldots, q_n) \in B^n$$

$$\mathcal{X}_{\mathcal{R}}(p_1, \ldots, p_n, q_1, \ldots, q_n) = 1 \Leftrightarrow \exists (M, M') \in \mathcal{R} \text{ s.t.}$$

$$\varepsilon(M) = (p_1, \ldots, p_n) \wedge \varepsilon(M') = (q_1, \ldots, q_n)$$

Given this, the elements of $\mathcal{M}$ that are in relation $\mathcal{R}$ with some elements of $\mathcal{M}'$ are defined by the set:

$$\mathcal{R}_M = \{M \in \mathcal{M} | \exists M', \text{ s.t. } (M, M') \in \mathcal{R}\}$$

Its characteristic function is defined as:

$$\mathcal{X}_{\mathcal{R}_M} = \exists_{q_1, \ldots, q_n} \mathcal{X}_{\mathcal{R}}(p_1, \ldots, p_n, q_1, \ldots, q_n)$$

3.3 EFFICIENT COMPUTATION OF A REACHABILITY SET

This section recalls the symbolic computation of the reachable markings of a Petri net by using BDDs, which is originally proposed by (Pastor *et al.*, 1994, 2001).

The set of markings at which transition t is enabled is defined as:

$$E_t = \prod_{p_i \in {}^\bullet t} p_i$$

The transition function for a transition $t \in T$ is a partially defined function $\delta^t: B^n \rightarrow B^n$ that transforms every marking $M \in E_t$ into a new marking $M' \in M_P$ by firing transition t. The function defines how the contents of each place are transformed by firing t at markings where t is enabled. The function is defined as $\forall i \in \{1, \ldots, n\}$,

$$\delta_i^t(p_1, \ldots, p_n) = \begin{cases} 1 & \text{if } p_i \in t^\bullet, \\ 0 & \text{if } p_i \in {}^\bullet t \text{ and } p_i \notin t^\bullet, \\ p_i & \text{otherwise.} \end{cases}$$

The function δ^t can be represented by a characteristic function $\mathcal{R}_t$ that requires two different sets of variables: $p_1, p_2, \ldots,$ and p_n for M and $q_1, q_2, \ldots,$ and q_n for M', respectively. According to the definition of δ^t, its characteristic function is described by the binary relation:

$$\mathcal{R}_t(p_1, \ldots, p_n, q_1, \ldots, q_n) = \prod_{i=1}^{n} (q_i \equiv \delta_i^t(p_1, \ldots, p_n)) \cdot E_t$$

Finding the set of markings that can be reached after firing transition t from any marking (at which t is enabled) in the set M is reduced to the computation of

$$Img(t, M) = \exists_{p_1, \ldots, p_n} [\mathcal{R}_t(p_1, \ldots, p_n, q_1, \ldots, q_n) \cdot M]$$

The union of the images of all transitions can be defined as:

$$Img(PN, M) = \exists_{p_1, \ldots, p_n} \sum_{\forall t_j \in T} \left[\prod_{i=1}^{n} \left(q_i \equiv \delta_i^{t_j}(p_1, \ldots, p_n) \right) \cdot E_t \cdot M \right]$$

This function can be used to calculate all markings that can be reached in one step from M. Given this, the set of reachable markings from M_0 can be calculated by using the image computation as follows:

Algorithm 3.1 (Pastor *et al.*, 2001) Computation of Reachable Markings
    Input: A safe Petri net system (N, M_0).
    Output: The set of reachable markings denoted as *Reach*.
    1) $From := Reach := \{M_0\}$.
    2) **repeat**
        $To := \emptyset$.
        foreach $\{t \in T\}$ **do**
            $To := To + Img(t, From)$.
        $New := To - Reach$.
        $From := New$.
        $Reach := Reach + New$.
        until $(New = \emptyset)$.
    3) Output *Reach*.
    4) End.

Now the Petri net analysis techniques by using BDDs can be extended to bounded nets with weights. A place $p \in P$ that contains up-to-k tokens can be represented by a set of Boolean variables, p^0, p^1, ..., and p^{k_p}. If the binary encoding scheme is used, the number of variables required for a k-bounded place is $\log_2(k+1)$. For example, a place p_1 containing five tokens can be represented by a vector of Boolean variables $p_1^2 p_1^1 p_1^0$.

To model the transition relations of a Petri net using binary variables, the weight of an arc is required to be encoded with the same type of encoding. The number of tokens in place p is represented by a set of Boolean variables p^0, p^1, ..., and p^{k_p} and weight $W(p,t)$ is represented by the same number of binary encoded Boolean constants ω^0, ω^1, ..., and ω^{k_p}. Then the relation $M(p) \geq W(p,t)$ can be represented as follows:

$$
\begin{aligned}
M(p) \geq W(p,t) \equiv & (p^{k_p} > \omega^{k_p}) \\
&+ (p^{k_p} \equiv \omega^{k_p}) \cdot (p^{k_p-1} > \omega^{k_p-1}) \\
&+ (p^{k_p} \equiv \omega^{k_p}) \cdot (p^{k_p-1} \equiv \omega^{k_p-1}) \cdot (p^{k_p-2} > \omega^{k_p-2}) \\
& \qquad \cdots \cdots \\
&+ (p^{k_p} \equiv \omega^{k_p}) \cdot (p^{k_p-1} \equiv \omega^{k_p-1}) \ldots (p^1 \equiv \omega^1) \cdot (p^0 > \omega^0) \\
&+ (p^{k_p} \equiv \omega^{k_p}) \cdot (p^{k_p-1} \equiv \omega^{k_p-1}) \ldots (p^1 \equiv \omega^1) \cdot (p^0 \equiv \omega^0)
\end{aligned}
$$

The markings at which a transition t is enabled can be defined by the characteristic function:

$$
E_t = \prod_{p \in {}^\bullet t} \left[\sum_{i=0}^{k_p} \left[(p^i > \omega^i) \cdot \prod_{j=i+1}^{k_p} (p^j \equiv \omega^j) \right] + \prod_{i=0}^{k_p} (p^i \equiv \omega^i) \right]
$$

In a place p_i, the fact that the content of the jth variable encoding the tokens is transformed after firing t can be represented by the function:

$$
\delta_i^{t^j}(p_1, \ldots, p_n) =
\begin{cases}
p_i^j \oplus \omega^j \oplus B_{j-1} & \text{if } p_i \in {}^\bullet t \setminus t^\bullet, \\
p_i^j \oplus \omega^j \oplus C_{j-1} & \text{if } p_i \in t^\bullet \setminus {}^\bullet t, \\
p_i^j \oplus \omega^j \oplus B_{j-1} & \text{if } p_i \in t^\bullet \cap {}^\bullet t \wedge W(p_i,t) > W(t,p_i), \\
p_i^j \oplus \omega^j \oplus C_{j-1} & \text{if } p_i \in t^\bullet \cap {}^\bullet t \wedge W(t,p_i) > W(p_i,t), \\
p_i^j & \text{otherwise};
\end{cases}
$$

where the carry and borrow functions are defined as:

$$
C_j =
\begin{cases}
p_i^j \cdot \omega^j + C_{j-1} \cdot (p_i^j \oplus \omega^j) & \text{if } j \geq 0, \\
0 & \text{otherwise,}
\end{cases}
$$

$$B_j = \begin{cases} \bar{p}_i^j \cdot \omega^j + B_{j-1} \cdot (p_i^j \equiv \omega^j) & \text{if } j \geq 0, \\ 0 & \text{otherwise.} \end{cases}$$

In each case, constant ω^j can be defined by the constant weighted value W, where W^j indicates the jth bit of W.

$$\omega^j = \begin{cases} W^j(p_i,t) & \text{if } p_i \in {}^{\bullet}t \backslash t^{\bullet}, \\ W^j(t,p_i) & \text{if } p_i \in t^{\bullet} \backslash {}^{\bullet}t, \\ [W(p_i,t) - W(t,p_i)]^j & \text{if } p_i \in t^{\bullet} \cap {}^{\bullet}t \wedge W(p_i,t) > W(t,p_i), \\ [W(t,p_i) - W(p_i,t)]^j & \text{if } p_i \in t^{\bullet} \cap {}^{\bullet}t \wedge W(t,p_i) > W(p_i,t). \end{cases}$$

Finally, the transition relation and image function can be extended to bounded and weighted nets are as follows:

$$\mathcal{R}_t(p_1,\ldots,p_n,q_1,\ldots,q_n) = \prod_{i=1}^{n} \prod_{j=0}^{k_{p_i}} (q_i^j \equiv \delta_i^{t^j}(p_1,\ldots,p_n))$$

$$Img(PN,\mathcal{M}) = \exists_{p_1,\ldots,p_n} \sum_{\forall t_j \in T} \left[\prod_{i=1}^{n} \prod_{j=0}^{k_{p_i}} (q_i \equiv \delta_i^{t^j}(p_1,\ldots,p_n)) \cdot E_t \cdot \mathcal{M} \right]$$

Now the set of reachable markings from M_0 can be calculated by Algorithm 3.1.

3.4 SYMBOLIC ANALYSIS OF A REACHABILITY GRAPH

This section presents a symbolic approach to the computation of legal markings and FBMs for bounded Petri nets by using BDDs.

3.4.1 Conversely Firing Policy

Let M and M' be two markings in $R(N,M_0)$. If $M[t\rangle M'$, then $M' = M + [N](\cdot,t)$ is true, and we have $M = M' - [N](\cdot,t)$. This property means that the predecessor markings can be calculated directly by reversing matrix $[N]$. Now we can define another transition firing policy, called a conversely firing policy as follows.

Definition 3.1 1) A transition t is conversely enabled at a marking M if $\forall p \in t^{\bullet}$: $M(p) \geq W(t,p)$. This fact is denoted by $M[-t\rangle$.

2) If $M[-t\rangle$, conversely firing it yields a new marking M'. It is denoted by $M[-t\rangle M'$. The new marking M' is defined as:

$$M'(p) = \begin{cases} M(p) + W(p,t) & p \in {}^{\bullet}t \backslash t^{\bullet}, \\ M(p) - W(t,p) & p \in t^{\bullet} \backslash {}^{\bullet}t, \\ M(p) + W(p,t) - W(t,p) & p \in {}^{\bullet}t \cap t^{\bullet}, \\ M(p) & \text{otherwise.} \end{cases}$$

3) The set of all markings that can be reached using the conversely firing policy from M_0 is called the co-reachability set. It is denoted as $R(-N, M_0)$.

This definition indicates that the co-reachability set has a property:

$$\forall M \in R(-N, M_0), M_0 \in R(N, M)$$

Note that M here is possibly not in $R(N, M_0)$. This property indicates that all the markings in the co-reachability set can reach M_0. In other words, $\forall M \in R(-N, M_0)$, there exists a firing sequence δ such that $M[\delta\rangle M_0$. The reachability set $R(N, M_0)$ and co-reachability set $R(-N, M_0)$ of the net model in Fig. 3.1 are shown in Fig. 3.2, which are represented in the form of directed graphs. It is easy to see that the marking $(0\ 0\ 1\ 1\ 0\ 0)^T$ is not reachable. However, by firing the sequence of transitions $t_4 t_3 t_4$, it reaches the initial marking $M_0 = (2\ 0\ 0\ 0\ 1\ 1)^T$.

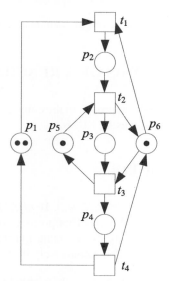

Figure 3.1 A Petri net example.

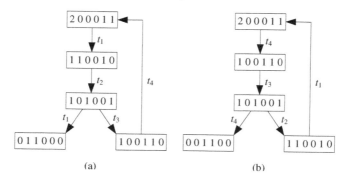

Figure 3.2 (a) the reachability set of Fig. 3.1 and (b) the co-reachability set of Fig. 3.1.

3.4.2 Efficient Computation of Legal Markings and FBMs

Given $R(-N, M_0)$, the set of legal markings can be calculated by the following expression due to the property of the co-reachability set.

$$\mathcal{M}_L = R(N, M_0) \cap R(-N, M_0) \tag{3.1}$$

Now we can develop a method to calculate the co-reachability set by using BDDs. First, we consider safe Petri nets. The policy to converse the firing of transition t can be defined as:

$$CImg(t, \mathcal{M}) = \prod_{i=1}^{n} (q_i \equiv \delta_i^{-t}(p_1, \dots, p_n)) \cdot E_{-t} \cdot \mathcal{M}$$

where

$$E_{-t} = \prod_{p_i \in t^\bullet} p_i,$$

$$\delta_i^{-t}(p_1, \dots, p_n) = \begin{cases} 1 & \text{if } p_i \in {}^\bullet t \text{ and } p_i \notin t^\bullet, \\ 0 & \text{if } p_i \in t^\bullet \text{ and } p_i \notin {}^\bullet t, \\ p_i & \text{otherwise.} \end{cases}$$

The set of markings $\mathcal{M}'$ that can be conversely reached after conversely firing all the transitions from all the markings in the set $\mathcal{M}$ (at which t is conversely enabled) is computed by the expression as follows:

$$CImg(PN, \mathcal{M}) = \sum_{\forall t_j \in T} \left[\prod_{i=1}^{n} (q_i \equiv \delta_i^{-t_j}(p_1, \dots, p_n)) \cdot E_{-t} \cdot \mathcal{M} \right]$$

Now we extend the conversely firing policy to weighted and bounded nets. We still use the binary encoding scheme. A place $p \in P$ that contains up-to-k tokens is represented by a set of Boolean variables, p^0, p^1, ..., and p^{k_p} and weight $W(t, p)$ is represented by the same number of binary encoded Boolean constants ω^0, ω^1, ..., and ω^{k_p}.

The markings at which a transition t is conversely enabled $(M(p) \geq W(t, p))$ are defined by the characteristic function E_{-t}:

$$\prod_{p \in t^\bullet} \left[\sum_{i=0}^{k_p} \left[(p^i > \omega^i) \cdot \prod_{j=i+1}^{k_p} (p^j \equiv \omega^j) \right] + \prod_{i=0}^{k_p} (p^i \equiv \omega^i) \right]$$

The transition function $\delta_i^{-t^j}$ for the jth variable encoding the tokens in a place p_i is

$$\delta_i^{-t^j}(p_1, \ldots, p_n) = \begin{cases} p_i^j \oplus \omega^j \oplus B_{j-1} & \text{if } p_i \in t^\bullet \setminus {}^\bullet t, \\ p_i^j \oplus \omega^j \oplus C_{j-1} & \text{if } p_i \in {}^\bullet t \setminus t^\bullet, \\ p_i^j \oplus \omega^j \oplus B_{j-1} & \text{if } p_i \in t^\bullet \cap {}^\bullet t \wedge W(t, p_i) > W(p_i, t), \\ p_i^j \oplus \omega^j \oplus C_{j-1} & \text{if } p_i \in t^\bullet \cap {}^\bullet t \wedge W(p_i, t) > W(t, p_i), \\ p_i^j & \text{otherwise;} \end{cases}$$

where the carry and borrow functions are defined as:

$$C_j = \begin{cases} p_i^j \cdot \omega^j + C_{j-1} \cdot (p_i^j \oplus \omega^j) & \text{if } j \geq 0, \\ 0 & \text{otherwise,} \end{cases}$$

$$B_j = \begin{cases} \overline{p_i^j} \cdot \omega^j + B_{j-1} \cdot (p_i^j \equiv \omega^j) & \text{if } j \geq 0, \\ 0 & \text{otherwise.} \end{cases}$$

In each case, constant ω^j can be defined by the constant weighted value W, where W^j indicates the jth bit of W.

$$\omega^j = \begin{cases} W^j(t, p_i) & \text{if } p_i \in t^\bullet \setminus {}^\bullet t, \\ W^j(p_i, t) & \text{if } p_i \in {}^\bullet t \setminus t^\bullet, \\ [W(t, p_i) - W(p_i, t)]^j & \text{if } p_i \in t^\bullet \cap {}^\bullet t \wedge W(t, p_i) > W(p_i, t), \\ [W(p_i, t) - W(t, p_i)]^j & \text{if } p_i \in t^\bullet \cap {}^\bullet t \wedge W(p_i, t) > W(t, p_i). \end{cases}$$

Finally, using the conversely firing policy, the transition relation and image function can be extended to bounded and weighted nets as follows:

$$\mathcal{R}_{-t}(p_1, \ldots, p_n, q_1, \ldots, q_n) = \prod_{i=1}^{n} \prod_{j=0}^{k_{p_i}} (q_i^j \equiv \delta_i^{-t^j}(p_1, \ldots, p_n))$$

$$CImg(PN, \mathcal{M}) = \exists_{p_1,\ldots,p_n} \sum_{\forall t_j \in T} \left[\prod_{i=1}^{n} \prod_{j=0}^{k_{p_i}} (q_i \equiv \delta_i^{-t_j}(p_1,\ldots,p_n)) \cdot E_{-t} \cdot \mathcal{M} \right]$$

Given this, for safe and bounded nets, we can use the same algorithms. The predecessor set of $\mathcal{M}$ can be found by function $CImg(PN, \mathcal{M})$. The co-reachablility set of the markings from M_0 can be computed by the algorithm as follows:

Algorithm 3.2 Computation of the Co-reachablility Set of Markings

Input: A Petri net system (N, M_0).
Output: The co-reachablility set of markings denoted as C_Reach.
1) $From:=C\_Reach:=\{M_0\}$.
2) **repeat**
    $To := \emptyset$.
    foreach $\{t \in T\}$ **do**
      $To := To + CImg(t, From)$.
    $New := To - Reach$.
    $From := New$.
    $C\_Reach := C\_Reach + New$.
    until $(New = \emptyset)$.
3) Output C_Reach.
4) End.

According to Eq. (3.1), the following algorithm is used to calculate the set of legal markings.

Algorithm 3.3 Computation of the Set of Legal Markings

Input: A Petri net system (N, M_0).
Output: The set of legal markings denoted as $\mathcal{M}_L$.
1) Compute reachablility set $Reach$ of the net system by Algorithm 3.1.
2) Compute co-reachablility set C_Reach of the net system by Algorithm 3.2.
3) $\mathcal{M}_L := Reach \cap C\_Reach$.
4) Output $\mathcal{M}_L$.
5) End.

Then the set of markings that should be removed from the reachable markings (represented by DZ) can be reduced as the following expression:

$$\mathcal{M}_{remove} = R(N, M_0) - \mathcal{M}_L$$

The successors of $\mathcal{M}_L$ can be computed by using function $Img(t, \mathcal{M}_L)$. For a transition t, if there exists a marking $M \in \mathcal{M}_L$ such that $M[t\rangle M'$ and $M' \notin \mathcal{M}_L$, then M' is an FBM. The following algorithm is used to compute the set of FBMs.

Algorithm 3.4 Computation of the Set of FBMs

Input: A Petri net system (N, M_0).

Output: The set of FBMs denoted as $\mathcal{M}_{\text{FBM}}$.
1) Compute the reachablility set *Reach* of the net system by Algorithm 3.1.
2) Compute the set of legal markings $\mathcal{M}_L$ of the net system by Algorithm 3.3.
3) $\mathcal{M}_{remove} := Reach - \mathcal{M}_L$.
4) $\mathcal{M}_{\text{FBM}} := \emptyset$.
5) **foreach** $\{t \in T\}$ **do**
    $New := Img(t, \mathcal{M}_L)$.
    $New := New \cap \mathcal{M}_{remove}$.
    if $\{New \neq \emptyset\}$ **then**
        $\mathcal{M}_{\text{FBM}} := \mathcal{M}_{\text{FBM}} + New$. /*The set of FBMs.*/
    endif
6) Output $\mathcal{M}_{\text{FBM}}$.
7) End.

3.4.3 Experimental Results

This section shows the efficiency of the algorithms of marking computation by an experimental study. Two scalable examples are used to illustrate the efficiency of the presented methods for large Petri nets. C++ program is used with the BDD tool Buddy-2.2 package (Lind-Nielsen, 2002) to implement the presented methods. The experimental study is conducted using a PC Intel CPU 2.80 GHz with 1 GByte RAM on the Linux operating system.

First, a well-known scalable dining philosophers problem is considered, as shown in Fig. 3.3. Table 3.1 shows the meanings of places and transitions. A certain number of philosophers spend their lives alternating between thinking and eating. They are seated around a circular table. There is a fork placed between each pair of neighboring philosophers. Each philosopher has access to the forks at his left and right. In order to eat, a philosopher must be in possession of both forks.

Table 3.1 Meanings of the places and transitions in the net in Fig. 3.3

Name	Meaning	Name	Meaning
p_{i_1}	idle	p_i	fork(resource place)
p_{i_2}	right hand empty	t_{i_1}	go to table
p_{i_3}	left hand empty	t_{i_2}	take right fork
p_{i_4}	taked right fork	t_{i_3}	take left fork
p_{i_5}	taked left fork	t_{i_4}	start eating
p_{i_6}	eating	t_{i_5}	leave forks and table

Table 3.2 shows the final experimental results which include the number of reachable markings (N_r), legal markings (N_l), FBMs (N_{FBM}), BDD nodes for reachable markings (N_r^{BDD}), legal markings (N_l^{BDD}), and FBMs ($N_{\text{FBM}}^{\text{BDD}}$), and finally the CPU time (in Seconds). The first column shows the size of the problem denoted by phil-*n*, where variable *n* represents the number of philosophers. The software

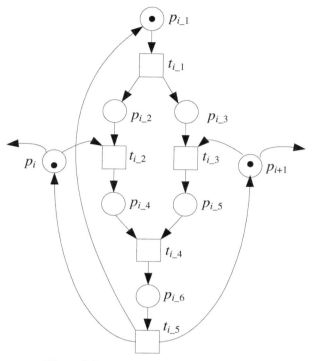

Figure 3.3 Petri net for a dining philosopher.

package INA2003 (Starke, 2003) can find the strongly connected components in the reachability graph of a Petri net. We hence show its CPU time (in Seconds) in the last column to do comparison.

Table 3.2 Results for scalable safe Petri nets of the Dining Philosophers Problem

Name	N_r	N_l	N_{FBM}	N_r^{BDD}	N_l^{BDD}	N_{FBM}^{BDD}	CPU	INA time
phil-5	2164	2162	2	248	280	66	<1	<1
phil-10	4.7×10^6	4.7×10^6	2	558	662	132	3	>12 hours
phil-15	1.0×10^{10}	1.0×10^{10}	2	864	1040	206	44	—
phil-20	2.2×10^{13}	2.2×10^{13}	2	1174	1420	276	237	—
phil-25	4.7×10^{16}	4.7×10^{16}	2	1484	1800	346	587	—
phil-30	1.0×10^{20}	1.0×10^{20}	2	1794	2180	416	1641	—
phil-35	2.2×10^{23}	2.2×10^{23}	2	2104	2560	486	6307	—

As shown in Table 3.2, we can see that when the number of reachable markings is not large, both BDDs and INA can finish the computation within a limited time. However, when the size of a Petri net gets too large, e.g., more than 10 philosophers, the finished time of the algorithm based on BDDs is still short but that of INA is so

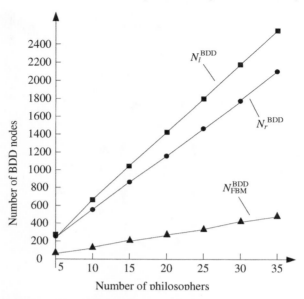

Figure 3.4 Number of BDD nodes for the Dining Philosophers Problem.

long that we cannot accept. Moreover, it is easy to see that the algorithm becomes more powerful with respect to INA with the increase of the size of the considered problem. Fig. 3.4 depicts the numbers of BDD nodes that represent the reachable markings, legal markings, and FBMs by using the fifth to seventh columns in Table 3.2, respectively. Clearly, the numbers of the reachable markings and legal markings grow very quickly, but the number of BDD nodes increases slightly, and, actually, linearly.

Now we consider bounded Petri nets. Fig. 3.5 shows a Petri net model of two sequential processes using shared resources. Table 3.3 presents the results of this Petri net model varying with the number of shared resources and the tokens of all the resources places $(R_1, R_2, \ldots,$ and $R_n)$ at the initial marking. For variables in the first column fmsld(n_1, n_2), n_1 represents the number of resources and n_2 represents the tokens in the resource places. We also let $M_0(p_{1\_0}) = M_0(p_{2\_0}) = n_2$. In the first four rows of the table, the nets have the same structure but the tokens of the resource places are different. In the first row, three bits are used to encode each place and in the second to fourth rows, four bits are used to encode each place. In the rest of the table, the nets have different resource places but every resource place has the same number of tokens at the initial marking. Two bits are used to encode each place since all places are 3-bounded.

From Table 3.3, it can be seen that when the size of the net model is small, BDDs have the similar efficiency as INA. However, if the net size becomes too large, INA needs much more time than BDDs. It can also be seen that the number of BDD nodes increases much more slowly than that of reachable markings with respect to the size of the Petri net model. However, the computational time still

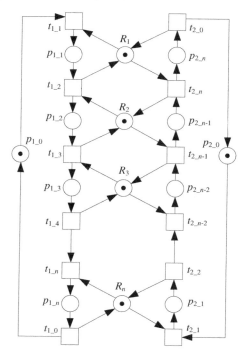

Figure 3.5 Petri net model of two sequential processes using shared resources.

Table 3.3 Results for scalable bounded Petri nets

Name	N_r	N_l	N_{FBM}	N_r^{BDD}	N_l^{BDD}	N_{FBM}^{BDD}	CPU	INA time
fmsld(4, 5)	13893	13890	3	22672	22621	112	11	8
fmsld(4, 10)	904758	904755	3	467952	467945	148	1595	>12hours
fmsld(4, 11)	1688190	1688187	3	698278	698271	150	5317	–
fmsld(4, 12)	3010069	3010066	3	1015563	1015556	146	7710	–
fmsld(5, 3)	2802	2798	4	6685	6679	115	3	<1
fmsld(6-3)	6505	6500	5	14988	14981	164	5	3
fmsld(7-3)	13554	13548	6	30749	30741	221	13	6
fmsld(8-3)	25994	25987	7	58785	58776	286	28	22
fmsld(9-3)	46682	43374	8	106096	106086	359	59	101
fmsld(10-3)	79477	79468	9	182525	182514	440	125	313
fmsld(11-3)	129450	129440	10	301538	301526	529	251	872
fmsld(12-3)	203114	203103	11	481134	481121	626	482	2222
fmsld(13-3)	308674	308662	12	744895	744881	731	886	5006

increases exponentially. Thus, BDDs are powerful to analyze the reachability graph of a Petri net and can overcome the state explosion problem to some degree but not completely.

3.5 EFFICIENT COMPUTATION OF MINIMAL SIPHONS

3.5.1 Symbolic Representation of Siphons

This section describes how siphons in Petri nets can be represented by using Boolean algebras. A siphon in a Petri net is a set of places $S = \{p_1, \ldots, p_k\}$. Let $\mathcal{S}_P = 2^P$ be the set of all subsets of the places representing siphons of a Petri net with $|P|$ places. The system $(2^{\mathcal{S}_P}, \cup, \cap, \emptyset, \mathcal{S}_P)$ is the Boolean algebra of the sets of siphons. It is isomorphic to the Boolean algebra of n-variable Boolean functions, where $n = |P|$ (see Theorem 2.3).

Given this, there is a one-to-one correspondence between siphons in $\mathcal{S}_P$ and vertices of B^n. Any siphon $S \in \mathcal{S}_P$ in a net can be represented by a vertex of B^n and determined by an encoding function $\varepsilon: \mathcal{S}_P \to B^n$. The image of every siphon $S \in \mathcal{S}_P$ is coded into a vertex $(p_1 \ldots p_n) \in B^n$ such that:

$$\forall i \in 1, \ldots, n, \quad p_i = \begin{cases} 1 & \text{if } p_i \in S, \\ 0 & \text{if } p_i \notin S. \end{cases} \tag{3.2}$$

As an example, both vertex $(0\ 1\ 0\ 0\ 0\ 1) \in B^6$ and cube $\bar{p}_1 p_2 \bar{p}_3 \bar{p}_4 \bar{p}_5 p_6$ represent a siphon S such that p_2 and p_6 are in S and p_1, p_3, p_4, and p_5 are not in S, i.e., $S = \{p_2, p_6\}$. Extending the use of the encoding function ε, each set of siphons $\mathcal{S} \in 2^{\mathcal{S}_P}$ has a corresponding image $V \in F_n(B)$ according to ε, defined by

$$V = \{v \in B^n | \exists S \in \mathcal{S} : v = \varepsilon(S)\} \tag{3.3}$$

Then the set $\mathcal{S}$ can be represented by those vertices for which function $\mathcal{X}_S: B^n \to B$ evaluates to 1, that is, $\mathcal{X}_S = \mathcal{X}_V$. For example, a set of siphons $\mathcal{S} = \{\{p_2, p_3\}, \{p_2, p_6\}\}$ has a characteristic function $\mathcal{X}_S$:

$$\mathcal{X}_S = \bar{p}_1 p_2 \bar{p}_4 \bar{p}_5 (p_3 \oplus p_6) \tag{3.4}$$

A characteristic function can also be used to represent a binary relation between two sets of siphons. Here, we need two sets of Boolean variables $p_1, p_2, \ldots$, and p_n and $q_1, q_2, \ldots$, and q_n to represent elements in two sets of siphons $\mathcal{S}$ and $\mathcal{S}'$, respectively. Let S and S' be any two siphons with $S \in \mathcal{S}$ and $S' \in \mathcal{S}'$. The relation $\mathcal{R} \subseteq \mathcal{S} \times \mathcal{S}'$ has a corresponding characteristic function:

$$\mathcal{X}_{\mathcal{R}}(p_1, \ldots, p_n, q_1, \ldots, q_n) = 1$$

$$\Leftrightarrow \exists (S, S') \in \mathcal{R} \text{ s.t. } \varepsilon(S) = (p_1, \ldots, p_n) \text{ and } \varepsilon(S') = (q_1, \ldots, q_n) \tag{3.5}$$

Thus, the elements of S that are in relation $\mathcal{R}$ with some elements of S' are defined as a set:

$$\mathcal{R}_S = \{S \mid S \in S, \exists S' \in S' \text{ s.t. } (S, S') \in \mathcal{R}\} \tag{3.6}$$

Its characteristic function is defined as:

$$\mathcal{X}_{\mathcal{R}_S} = \exists_{q_1,\ldots,q_n} \mathcal{X}_{\mathcal{R}}(p_1,\ldots,p_n,q_1,\ldots,q_n) \tag{3.7}$$

The relation function is used to identify non-minimal siphons from all siphons in a Petri net. Then, the minimal siphons can be found by removing all non-minimal ones, as reported in the following section.

3.5.2 Symbolic Extraction of Minimal Siphons

This section presents a symbolic method to compute the siphons and minimal siphons in a Petri net.

Let S be a siphon of a Petri net N with $N = (P, T, F, W)$. According to the definition of siphons, if a place $p_i \in S$, then $\forall t \in {}^\bullet p_i$, $\exists p_j \in S$ s.t. $t \in p_j^\bullet$. Thus, the characteristic function $\mathcal{X}_{p_i}$ of place p_i in the sense of siphons is defined as

$$\mathcal{X}_{p_i} = \overline{p_i} + \prod_{t \in {}^\bullet p_i} \sum_{p_j \in {}^\bullet t} p_j \tag{3.8}$$

For function $\mathcal{X}_{p_i}$, we have the following two lemmas to show its relation with siphons in net N.

Lemma 3.1 *Let S be a set of places and v_S be the corresponding vertex in B^n that represents S. The fact that function $\mathcal{X}_{p_i}$ evaluates to 1 for v_S indicates that if a place $p_i \in S$, then $\forall t \in {}^\bullet p_i$, $\exists p_j \in S$ s.t. $t \in p_j^\bullet$.*

Proof If $p_i \in S$, then $p_i = 1$ in v_S. Thus, we have $\bar{p}_i = 0$. In this case, the fact that function $\overline{p_i} + \prod_{t \in {}^\bullet p_i} \sum_{p_j \in {}^\bullet t} p_j$ evaluates to 1 for v_S indicates that function $\prod_{t \in {}^\bullet p_i} \sum_{p_j \in {}^\bullet t} p_j$ must evaluate to 1 for v_S. Therefore, for any $t \in {}^\bullet p_i$, function $\sum_{p_j \in {}^\bullet t} p_j$ must evaluate to 1 for v_S, implying that at least one place variable $p_j \in {}^\bullet t$ must evaluate to 1 in v_S, i.e., $\exists p_j \in S$. As a result, for any $t \in {}^\bullet p_i$, we have $\exists p_j \in S$ s.t. $t \in p_j^\bullet$. □

Lemma 3.2 *Let S be a siphon and v_S be the corresponding vertex in B^n that represents S. Function $\mathcal{X}_{p_i}$ evaluates to 1 for v_S.*

Proof There are two cases to be considered.

1. If $p_i \notin S$, then $p_i = 0$ in v_S. Thus, $\bar{p}_i$ evaluates to 1 for v_S. It is easy to see that function $\mathcal{X}_{p_i}$ evaluates to 1 for v_S.
2. If $p_i \in S$, then $p_i = 1$ in v_S. $\forall t \in {}^\bullet p_i$, we have $t \in {}^\bullet S$. Since S is a siphon, $t \in S^\bullet$ is true. That is to say, $\exists p_j \in S$ s.t. $p_j \in {}^\bullet t$. Thus, $\sum_{p_j \in {}^\bullet t} p_j$ evaluates to 1 for

v_S and the conjunction $\prod_{t \in {}^\bullet p_i} \sum_{p_j \in {}^\bullet t} p_j$ evaluates to 1 for v_S. We conclude that function $\mathcal{X}_{p_i}$ evaluates to 1 for v_S.   □

By summarizing the characteristic functions of all places in P, we have the characteristic function $\mathcal{X}_{\mathcal{S}_N}$ of the set $\mathcal{S}_N$ of all siphons in net N:

$$\mathcal{X}_{\mathcal{S}_N} = \prod_{p_i \in P} \mathcal{X}_{p_i} \tag{3.9}$$

Obviously, $\mathcal{S}_N$ includes the empty set. Removing the empty set from $\mathcal{S}_N$, we have the characteristic function $\mathcal{X}_S$ that represents the set S of all the siphons in net N as follows:

$$\mathcal{X}_S = \prod_{p_i \in P} \mathcal{X}_{p_i} \cdot \sum_{p_j \in P} p_j \tag{3.10}$$

Theorem 3.1 *Function $\mathcal{X}_S$ represents all the siphons in net N.*

Proof First we claim that any set of places represented by $\mathcal{X}_S$ is a siphon. Let v_S be any vertex of B^n for which $\mathcal{X}_S$ evaluates to 1 and S a set of places represented by v_S. In function $\mathcal{X}_S$, $\sum_{p_j \in P} p_j$ ensures that there is at least one place in S, i.e., $S \neq \emptyset$. According to Lemma 3.1, $\prod_{p_i \in P} \mathcal{X}_{p_i}$ ensures that $\forall t \in {}^\bullet p_i$, $\exists p_j \in S$ s.t. $p_j \in t^\bullet$, i.e., ${}^\bullet S \subseteq S^\bullet$. Therefore, S is a siphon.

Now we show that for any siphon S in N, $\mathcal{X}_S$ evaluates to 1 for its corresponding vertex v_S in B^n. Since $S \neq \emptyset$, $\sum_{p_j \in P} p_j$ evaluates to 1 for v_S. According to Lemma 3.2, since S is a siphon, $\forall p_i \in P$, $\mathcal{X}_{p_i}$ evaluates to 1 for v_S. Thus, the conjunction $\prod_{p_i \in P} \mathcal{X}_{p_i}$ evaluates to 1 for v_S. Finally, $\mathcal{X}_S$ evaluates to 1 for its corresponding vertex v_S. Function $\mathcal{X}_S$ represents all siphons in net N.   □

For large-scale Petri nets, though the number of siphons is huge, experimental results show that the number of nodes of the BDD representing $\mathcal{X}_S$ is much smaller than that of the siphons and the computation for $\mathcal{X}_S$ can be finished within a reasonable time.

Now we consider how to extract the minimal siphons from S. Here, we need two sets of Boolean variables $p_1, p_2, \ldots,$ and p_n and $q_1, q_2, \ldots,$ and q_n to represent the elements in two sets S and S', respectively. A binary relation $\mathcal{R}_\subset \subseteq S \times S'$ is defined as $(S, S') \in \mathcal{R}_\subset \Leftrightarrow S \subset S'$. Then, its characteristic function $\mathcal{X}_{\mathcal{R}_\subset}$ can be defined as follows:

$$\mathcal{X}_{\mathcal{R}_\subset}(p_1, \ldots, p_n, q_1, \ldots, q_n) = \prod_{i=1}^{n}(\bar{p}_i + q_i) \cdot \sum_{i=1}^{n}(p_i \oplus q_i) \tag{3.11}$$

Lemma 3.3 *Let S and S' be two siphons that are represented by two sets of Boolean variables $p_1, p_2, \ldots,$ and p_n and $q_1, q_2, \ldots,$ and q_n, respectively. v_S and $v_{S'}$ are two corresponding vertices of S and S', respectively. The fact that function $\mathcal{X}_{\mathcal{R}_\subset}(p_1, \ldots, p_n, q_1, \ldots, q_n)$ evaluates to 1 for v_S and $v_{S'}$ indicates that $S \subset S'$.*

Proof The fact that function $\mathcal{X}_{\mathcal{R}_\subset}(p_1, \ldots, p_n, q_1, \ldots, q_n)$ evaluates to 1 for v_S and $v_{S'}$ indicates that the conjunction $\prod_{i=1}^{n}(\bar{p}_i + q_i)$ and $\sum_{i=1}^{n}(p_i \oplus q_i)$ must evaluate to 1 for v_S and $v_{S'}$. First, the fact that $\prod_{i=1}^{n}(\bar{p}_i + q_i)$ evaluates to 1 for v_S and $v_{S'}$ means that $\forall i \in \{1, \ldots, n\}$, 1) if $q_i = 0$ in $v_{S'}$, then $p_i = 0$ in v_S; and 2) if $q_i = 1$ in $v_{S'}$, then $p_i = 1$ or $p_i = 0$ in v_S. Therefore, we have $S \subseteq S'$. Second, the fact that $\sum_{i=1}^{n}(p_i \oplus q_i)$ evaluates to 1 for v_S and $v_{S'}$ means that there is at least one pair of places p_i and q_i satisfying that $p_i \oplus q_i$ evaluates to 1 for v_S and $v_{S'}$, i.e., $p_i \neq q_i$. Thus, we have $S \neq S'$. Finally, we have $S \subset S'$. □

This relation function can also be used to represent the relation between two sets of siphons. Let $\mathcal{R}_{\supset S}$ be a subset of $\mathcal{S}$ such that every siphon in $\mathcal{R}_{\supset S}$ strictly includes another siphon in $\mathcal{S}$, which is defined as follows:

$$\mathcal{R}_{\supset S} = \{S \in \mathcal{S} | \exists S' \in \mathcal{S}, \text{ s.t. } S' \subset S\} \tag{3.12}$$

Using the relation function $\mathcal{X}_{\mathcal{R}_{\subset}}$, the characteristic function of $\mathcal{R}_{\supset S}$ can be defined as follows:

$$\mathcal{X}_{\mathcal{R}_{\supset S}} = \exists_{q_1, \ldots, q_n} \mathcal{R}_{\subset}(p_1, \ldots, p_n, q_1, \ldots, q_n) \cdot \mathcal{X}_S \cdot \mathcal{X}_{S'} \tag{3.13}$$

where $\mathcal{X}_{S'}$ is obtained from function $\mathcal{X}_S$ via replacing each variable p_i by q_i, implying that $\mathcal{X}_{S'}$ and $\mathcal{X}_S$ represent the same set S with different sets of variables.

Lemma 3.4 *Each siphon in the set $\mathcal{R}_{\supset S}$ represented by function $\mathcal{X}_{\mathcal{R}_{\supset S}}$ has a siphon in $\mathcal{S}$ as a proper subset.*

Proof Let S and S' be two sets of places with corresponding vertices $v_S = (p_1, \ldots, p_n)$ and $v_{S'} = (q_1, \ldots, q_n)$, respectively, for which function $\mathcal{R}_{\subset}(p_1, \ldots, p_n, q_1, \ldots, q_n) \cdot \mathcal{X}_S \cdot \mathcal{X}_{S'}$ evaluates to 1. Thus, $\mathcal{X}_S$ evaluates to 1 for v_S and $\mathcal{X}_{S'}$ evaluates to 1 for $v_{S'}$. According to Theorem 3.1, we have $S \in \mathcal{S}$ and $S' \in \mathcal{S}$. Since $\mathcal{R}_{\subset}(p_1, \ldots, p_n, q_1, \ldots, q_n)$ evaluates to 1 for v_S and $v_{S'}$, by Lemma 3.3, we have $S \subset S'$. In Eq. (3.13), since the vertex with variables $q_1, q_2, \ldots$, and q_n is selected as the vertex of $\mathcal{X}_{\mathcal{R}_{\supset S}}$, i.e., $v_{S'}$, S' is in $\mathcal{R}_{\supset S}$. Finally, each siphon in $\mathcal{R}_{\supset S}$ has a siphon in $\mathcal{S}$ as a proper subset. □

The characteristic function $\mathcal{X}_{S_{MS}}$ of the set S_{MS} of minimal siphons can be calculated as follows:

$$\mathcal{X}_{S_{MS}} = \mathcal{X}_S - \mathcal{X}_{\mathcal{R}_{\supset S}} \tag{3.14}$$

Theorem 3.2 *Function $\mathcal{X}_{S_{MS}}$ represents all minimal siphons in net N.*

Proof According to Lemma 3.4, any siphon in $\mathcal{S}$ that has a proper subset is in $\mathcal{R}_{\supset S}$. Thus, only the ones that do not contain other siphons in $\mathcal{S}$ as a proper subset are preserved. That is to say, function $\mathcal{X}_{S_{MS}}$ represents all the minimal siphons in net N. □

$\mathcal{X}_{S_{MS}}$ is the function that evaluates to 1 for those vertices corresponding to all minimal siphons. Thus, once we obtain $\mathcal{X}_{S_{MS}}$, the minimal siphons of a Petri net can be easily identified.

3.5.3 An Illustrative Example

This section presents an example to show the application of BDDs to compute the minimal siphons in a Petri net.

A Petri net is shown in Fig. 3.6. It has six places and four transitions. Six Boolean variables p_1, p_2, p_3, p_4, p_5, and p_6 are used to represent the six places.

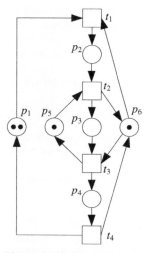

Figure 3.6 A Petri net example.

In Fig. 3.6, we have ${}^\bullet p_1 = \{t_4\}$ and ${}^\bullet t_4 = \{p_4\}$. Thus, if p_1 is in a siphon, p_4 must be in the siphon. The characteristic function for p_1 is defined as follows:

$$X_{p_1} = \overline{p}_1 + p_4$$

Similarly, for the other places, we have

$$X_{p_2} = \overline{p}_2 + p_1 + p_6$$
$$X_{p_3} = \overline{p}_3 + p_2 + p_5$$
$$X_{p_4} = \overline{p}_4 + p_3 + p_6$$
$$X_{p_5} = \overline{p}_5 + p_3 + p_6$$
$$X_{p_6} = \overline{p}_6 + (p_2 + p_5) \cdot p_4$$

Summarizing the characteristic functions of all places, we have the characteristic function X_{S_N}:

$$X_{S_N} = X_{p_1} \cdot X_{p_2} \cdot X_{p_3} \cdot X_{p_4} \cdot X_{p_5} \cdot X_{p_6}$$

Removing the empty set, the characteristic function of the set of all siphons is presented as follows:

$$\mathcal{X}_S = \mathcal{X}_{S_N} \cdot (p_1 + p_2 + p_3 + p_4 + p_5 + p_6)$$

$$= \bar{p}_1 \bar{p}_2 \bar{p}_3 p_4 p_5 p_6 + \bar{p}_1 \bar{p}_2 p_3 \bar{p}_4 p_5 \bar{p}_6 + \bar{p}_1 \bar{p}_2 p_3 p_4 p_5 + \bar{p}_1 p_2 p_4 p_6$$

$$+ p_1 \bar{p}_2 \bar{p}_3 p_4 p_5 p_6 + p_1 \bar{p}_2 p_3 p_4 p_5 + p_1 p_2 \bar{p}_3 p_4 p_6 + p_1 p_2 p_3 p_4 \quad (3.15)$$

Using the order $p_1 < p_2 < p_3 < p_4 < p_5 < p_6$, $\mathcal{X}_S$ can be represented by a BDD with 16 nodes, as shown in Fig. 3.7. It has 17 truth vertices, which indicates that the Petri net has total 17 siphons. Each of the 17 siphons can be identified easily from the final expression of $\mathcal{X}_S$ or the BDD structure in Fig. 3.7. For example, in expression (3.15), $\bar{p}_1 \bar{p}_2 \bar{p}_3 p_4 p_5 p_6$ represents a siphon $\{p_4, p_5, p_6\}$ and $\bar{p}_1 \bar{p}_2 p_3 p_4 p_5$ represents two siphons $\{p_3, p_4, p_5\}$ and $\{p_3, p_4, p_5, p_6\}$. All the 17 siphons are shown in Table 3.4. In Fig. 3.7, there are 17 paths from the root node to the sink node labeled with 1, where each path has a one-to-one corresponding siphon.

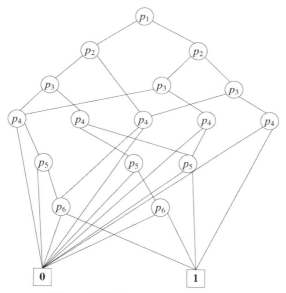

Figure 3.7 BDD representation of $\mathcal{X}_S$.

Now, we identify the minimal siphons from all the 17 siphons. Another set of variables $q_1, q_2, \ldots,$ and q_6 are used to represent the siphons in a new set $\mathcal{S}'$. Let S and S' be two siphons with $S \in \mathcal{S}$ and $S' \in \mathcal{S}'$. The characteristic function of relation $S \subset S'$ is defined as follows:

$$\mathcal{X}_{\mathcal{R}_\subset}(p_1, \ldots, p_6, q_1, \ldots, q_6) = \prod_{i=1}^{6}(\bar{p}_i + q_i) \cdot \sum_{i=1}^{6}(p_i \oplus q_i)$$

The characteristic function $\mathcal{X}_{\mathcal{R}_{\supset S}}$ of the set $\mathcal{R}_{\supset S}$ such that every siphon in $\mathcal{R}_{\supset S}$ strictly includes another siphon in $\mathcal{S}$ is calculated as follows:

$$\mathcal{X}_{\mathcal{R}_{\supset S}} = \exists_{q_1, \ldots, q_6} \mathcal{X}_{\mathcal{R}_\subset}(p_1, \ldots, p_6, q_1, \ldots, q_6) \cdot \mathcal{X}_S \cdot \mathcal{X}_{S'}$$

Table 3.4 The 17 siphons represented by $\mathcal{X}_S$ in Fig. 3.7

BDD representation	Siphons
$\bar{p}_1\bar{p}_2\bar{p}_3 p_4 p_5 p_6$	$\{p_4,p_5,p_6\}$
$\bar{p}_1\bar{p}_2 p_3 \bar{p}_4 p_5 \bar{p}_6$	$\{p_3,p_5\}$
$\bar{p}_1\bar{p}_2 p_3 p_4 p_5$	$\{p_3,p_4,p_5\}, \{p_3,p_4,p_5,p_6\}$
$\bar{p}_1 p_2 p_4 p_6$	$\{p_2,p_4,p_6\}, \{p_2,p_3,p_4,p_6\}, \{p_2,p_4,p_5,p_6\}, \{p_2,p_3,p_4,p_5,p_6\}$
$p_1\bar{p}_2\bar{p}_3 p_4 p_5 p_6$	$\{p_1,p_4,p_5,p_6\}$
$p_1\bar{p}_2 p_3 p_4 p_5$	$\{p_1,p_3,p_4,p_5\}, \{p_1,p_3,p_4,p_5,p_6\}$
$p_1 p_2\bar{p}_3 p_4 p_6$	$\{p_1,p_2,p_4,p_6\}, \{p_1,p_2,p_4,p_5,p_6\}$
$p_1 p_2 p_3 p_4$	$\{p_1,p_2,p_3,p_4\}, \{p_1,p_2,p_3,p_4,p_5\}, \{p_1,p_2,p_3,p_4,p_6\},$ $\{p_1,p_2,p_3,p_4,p_5,p_6\}$

where $\mathcal{X}_{S'}$ is obtained from function $\mathcal{X}_S$ via replacing each variable p_i by q_i. Then, the characteristic function $\mathcal{X}_{S_{MS}}$ of the set $\mathcal{S}_{MS}$ of minimal siphons can be calculated as follows:

$$\mathcal{X}_{S_{MS}} = \mathcal{X}_S - \mathcal{X}_{\mathcal{R}_{\supset S}}$$
$$= \bar{p}_1\bar{p}_2\bar{p}_3 p_4 p_5 p_6 + \bar{p}_1\bar{p}_2 p_3 p_4 p_5 \bar{p}_6 + \bar{p}_1 p_2\bar{p}_3 p_4\bar{p}_5 p_6 + p_1 p_2 p_3 p_4\bar{p}_5\bar{p}_6$$

By using the order $p_1 < p_2 < p_3 < p_4 < p_5 < p_6$, $\mathcal{X}_{S_{MS}}$ can be represented by a BDD with 16 nodes, as shown in Fig. 3.8. It has four truth vertices, implying that the Petri net has totally four minimal siphons, i.e., $\{p_4,p_5,p_6\}$, $\{p_3,p_5\}$, $\{p_2,p_4,p_6\}$, and $\{p_1,p_2,p_3,p_4\}$.

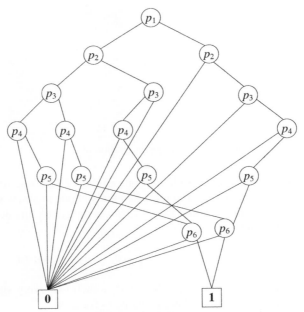

Figure 3.8 BDD representation of $\mathcal{X}_{MS}$.

3.5.4 Experimental Results

This subsection provides experimental results of the presented method. For the application of BDDs, C++ program is used with the BDD tool Buddy-2.2 package (Lind-Nielsen, 2002) on a Linux operating system with Intel CPU Core 2.8 GHz and 4 GB memory.

First, the well-known scalable dining philosophers problem in Fig. 3.3 is considered. As shown in Table 3.5, the final experimental results include the number of places $|P|$, the number of transitions $|T|$, the number of all siphons N_S, the number of minimal siphons N_{MS}, the number of nodes of the BDD that represents all the siphons N_S^{BDD}, the number of nodes of the BDD that represents the minimal siphons N_{MS}^{BDD}, and the CPU time (in CPU Seconds). The first column shows the size of the problem denoted by phi-n, where variable n represents the number of philosophers.

Table 3.5 Results for scalable Petri nets of the dining philosophers problem

name	$|P|$	$|T|$	N_S	N_{MS}	N_S^{BDD}	N_{MS}^{BDD}	CPU time (S)
phi-10	70	50	1.4×10^{13}	32	933	463	< 1
phi-50	350	250	4.5×10^{65}	152	5285	2544	3
phi-100	700	500	2.0×10^{131}	302	10735	5144	15
phi-150	1050	750	9.1×10^{196}	452	16185	7744	52
phi-200	1400	1000	4.1×10^{262}	602	21635	10344	124
phi-250	1750	1250	1.9×10^{328}	752	27085	12944	251
phi-300	2100	1500	8.3×10^{393}	902	32535	15544	439
phi-350	2450	1750	3.8×10^{459}	1052	37985	18144	1197
phi-400	2800	2000	1.7×10^{525}	1202	43435	20744	1740
phi-450	3150	2250	7.6×10^{590}	1352	48885	23344	2312
phi-500	3500	2500	3.4×10^{656}	1502	54335	25944	3284
phi-550	3850	2750	1.5×10^{722}	1652	59785	28544	4783
phi-600	4200	3000	7.0×10^{787}	1802	65235	31144	6067

As shown in Table 3.5, we can see that the presented method can compute rather efficiently the minimal siphons in the Petri net models of the dining philosophers problem. For the last row, even though the net model has as many as 4,200 places and 3,000 transitions, the computation can be finished in less than two hours. The number of siphons in a Petri net increases rapidly with its size. For large-scale Petri nets, it is a huge number (see the fourth column in Table 3.5). However, the number of the nodes of the BDD that represents the siphons is much smaller than that of siphons. In fact, it increases slightly, and actually linearly .

Now we consider a number of randomly generated Petri nets with different sizes and topology. The computational results are presented in Tables 3.6, 3.7, and 3.8, with $d_i = d_o = 0.2$, $d_i = d_o = 0.5$, and $d_i = d_o = 0.8$, respectively, where d_i and d_o are the probability densities of input and output arcs to all possible combinations, respectively. The software package INA2003 (Starke, 2003) can also compute the minimal siphons in a Petri net. We hence show its CPU time (in seconds) in the last column of the three tables for the comparison purposes.

Table 3.6 Results for randomly generated Petri nets with $d_i = d_o = 0.2$

$\lvert P \rvert$	$\lvert T \rvert$	N_S	N_{MS}	N_S^{BDD}	N_{MS}^{BDD}	CPU time (S)	INA time (S)
5	5	0	0	0	0	< 1	< 1
10	10	3	2	16	17	< 1	< 1
15	15	6094	19	179	91	< 1	< 1
20	20	1840	35	266	167	< 1	4
25	25	4.0×10^6	261	2124	1054	< 1	60
30	30	5.3×10^8	11724	40490	20329	25	>7200
35	35	8.3×10^9	29335	107048	48530	303	—
40	40	5.6×10^{11}	—	479104	—	—	—

Table 3.7 Results for randomly generated Petri nets with $d_i = d_o = 0.5$

$\lvert P \rvert$	$\lvert T \rvert$	N_S	N_{MS}	N_S^{BDD}	N_{MS}^{BDD}	CPU time (S)	INA time (S)
5	5	14	3	6	9	< 1	< 1
10	10	702	6	26	35	< 1	< 1
15	15	28902	116	279	228	< 1	< 1
20	20	1.0×10^6	566	916	899	< 1	2
25	25	3.3×10^7	2271	4043	3367	2	12
30	30	1.1×10^9	7313	9791	9908	13	90
35	35	3.4×10^{10}	19776	31085	26576	85	680
40	40	1.1×10^{12}	56127	60716	67248	523	>7200
45	45	3.5×10^{13}	—	176728	—	—	—

Table 3.8 Results for randomly generated Petri nets with $d_i = d_o = 0.8$

$\lvert P \rvert$	$\lvert T \rvert$	N_S	N_{MS}	N_S^{BDD}	N_{MS}^{BDD}	CPU time (S)	INA time (S)
5	5	14	3	6	9	< 1	< 1
10	10	964	18	32	42	< 1	< 1
15	15	32591	60	91	99	< 1	< 1
20	20	1.0×10^6	124	144	223	< 1	< 1
25	25	3.4×10^7	355	459	551	< 1	< 1
30	30	1.1×10^9	839	1034	1179	< 1	1
35	35	3.4×10^{10}	1770	2052	2686	1	2
40	40	1.1×10^{12}	3114	3073	4353	3	5
45	45	3.5×10^{13}	4719	5901	7352	5	12
50	50	1.1×10^{14}	7411	9559	12129	11	31
55	55	3.6×10^{16}	12048	15011	19917	30	70
60	60	1.2×10^{17}	16017	21343	25854	58	100
65	65	3.6×10^{19}	24349	25602	37178	108	311
70	70	1.2×10^{21}	36142	47679	61155	248	603
75	75	3.8×10^{22}	49466	58182	82069	407	1261
80	80	1.2×10^{24}	65596	88724	121152	920	4860
85	85	3.9×10^{25}	90666	108882	165752	2212	>7200

For the results in the three tables, we reach the following conclusions:

1. In the last row of Table 3.6, all the siphons can be found within nine seconds. However, identification of minimal siphons from the siphons is interrupted since it cannot be finished within two hours. Similarly, in the last row of Table 3.7, all siphons can be found in ten seconds and the computation for finding minimal siphons cannot be finished. Thus, it can be seen that the computation of siphons in Petri nets by using BDDs is very efficient but extracting minimal siphons from all the siphons requires much more time.
2. The bigger d_i and d_o become, the shorter time the presented method requires. Hence, the larger sized Petri nets can be dealt with.
3. As far as the BDD nodes are concerned, it can be seen that for large sets of encoded data (the sets of siphons), the number of BDD nodes is much smaller than that of siphons that are represented by the BDDs. On the other hand, for the minimal siphons, their number is more than that of BDD nodes. This shows that BDDs are capable of representing large sets of data with a small data structure. However, for some small sets of minimal siphons, BDDs may have more nodes than the number of minimal siphons.
4. From the viewpoint of computational time, it is easy to see that the presented method becomes more efficient with respect to INA with the increase of the size of the plant.

Note that Cordone *et al.* (2005) provide very similar results for randomly generated Petri nets in the sense of computational time.

Finally, we present some results by applying the reported method to deal with the Petri net models in the literature. Most of the nets presented in the literature are not very big. Thus, the method can deal with them in a very short time. For example, for the classical net model in (Ezpeleta *et al.*, 1995) with 26 places and 20 transitions, the method can find all 28 minimal siphons in less than one second. For this net, both the method and INA can enumerate the minimal siphons in a very short time. A large-scale Petri net model with 48 places and 38 transitions can be found in (Li and Zhou, 2008). For the net model, the method can find 5.5×10^{10} siphons and 87 minimal siphons in 30 seconds. If we use INA to deal with the net model, it requires 463 seconds to compute the minimal siphons. Thus, the method is also very efficient to deal with the Petri net models in the literature.

3.6 CONCLUSIONS

Reachability graphs and siphons play important roles in the deadlock control of Petri nets. This chapter presents symbolic approaches to find all reachable markings and minimal siphons of Petri nets by using BDDs. This chapter also provides an algorithm to find the set of legal markings and the set of FBMs. The methods are general and can be applied to all classes of Petri nets. BDDs have the capability of representing large sets of data with small shared data structures. All computation in the presented methods is implemented by BDDs. Experimental results show their efficiency via studying a number of typical examples.

The numbers of both reachable markings and siphons in a Petri net increase exponentially with respect to its size, which can be verified by the presented experimental results. BDDs can be used to handle the problems efficiently. The experimental results show that the computational time for finding all siphons can be finished in a very short time but identifying minimal siphons from all siphons requires much more time. Thus, deciding how to accelerate BDD operations for identifying minimal siphons is still under research.

3.7 BIBLIOGRAPHICAL REMARKS

This chapter provides the applications of BDDs for the reachability graph analysis and structural analysis of Petri nets. Seminal work on BDD-based computation of all reachable markings can be found in (Pastor *et al.*, 1994, 2001). The most material for the analysis of reachability graphs by using BDDs can be found in (Chen *et al.*, 2011). For the computation of the minimal siphons in a Petri net, the readers are encouraged to refer to (Chen and Liu, 2013) for details. Also, other efficient methods for the computation of siphons can be found in (Chao, 2006a,b; Cordone *et al.*, 2005; Ezpeleta *et al.*, 1993; Wang *et al.*, 2009). There are still other data decision structures that can be used to deal with a large set of data such as data decision diagrams (DDDs) (Couvreur *et al.*, 2002), multi-valued decision diagrams (MDDs) (Miner and Ciardo, 1999), and zero-suppressed binary decision diagrams (ZBDDs) (Minato, 2000). A very popular tool to compute siphons and reachability graphs of Petri nets is INA (Starke, 2003).

References

Andersen, H. R. 1997. An introduction to binary decision diagrams. Lecture Notes for 49285 Advanced Algorithms E97, Department of Information Technology, Technical University of Denmark.

Brant, R. 1992. Symbolic Boolean manipulation with ordered binary decision diagrams. ACM Computing Surveys. 24(3): 293–318.

Chao, D. Y. 2006a. Computation of elementary siphons in Petri nets for deadlock control. Computer Journal. 49(4): 470–479.

Chao, D. Y. 2006b. Searching strict minimal siphons for SNC-based resource allocation systems. Journal of Information Science and Engineering. 23(3): 853–867.

Chen, Y. F., Z. W. Li, M. Khalgui, and O. Mosbahi. 2011. Design of a maximally permissive liveness-enforcing Petri net supervisor for flexible manufacturing systems. IEEE Transactions on Automation Science and Engineering. 8(2): 374–393.

Chen, Y. F. and G. Y. Liu. 2013. Computation of minimal siphons in Petri nets by using binary decision diagrams. ACM Transactions on Embedded Computing Systems. 12(1).

Cordone, R., L. Ferrarini, and L. Piroddi. 2005. Enumeration algorithms for minimal siphons in Petri nets based on place constraints. IEEE Transactions on System, Man and Cybernetics, Part A. 35(6): 844–854.

Couvreur, J. M., E. Encrenaz, E. Paviot-Adet, D. Poitrenaud, and P. A. Wacrenier. 2002. Data decision diagrams for Petri net analysis. Lecture Notes in Computer Science. 2360: 129–158.

Ezpeleta, J., J. M. Couvreur, and M. Silva. 1993. A new technique for finding a generating family of siphons, traps, and st-components: Application to colored Petri nets. Lecture Notes in Computer Science. 674: 126–147.

Ezpeleta, J., J. M. Colom, and J. Martinez. 1995. A Petri net based deadlock prevention policy for flexible manufacturing systems. IEEE Transactions on Robotics and Automation. 11(2): 173–184.

Ezpeleta, J. and L. Recalde. 2004. A deadlock avoidance approach for nonsequential resource allocation systems. IEEE Transactions on Systems, Man, and Cybernetics, Part A. 34(1): 93–101.

Ghaffari, A., N. Rezg, and X. L. Xie. 2003. Design of a live and maximally permissive Petri net controller using the theory of regions. IEEE Transactions on Robotics and Automation. 19(1): 137–142.

Huang, Y. S., M. D. Jeng, X. L. Xie, and S. L. Chung. 2001. Deadlock prevention based on Petri nets and siphons. International Journal of Production Research. 39(2): 283–305.

Jeng, M. D. and X. L. Xie. Deadlock detection and prevention of automated manufacturing systems using Petri nets and siphons. pp. 233–281. In M. C. Zhou and M. P. Fanti. [eds.]. 2005. Deadlock Resolution in Computer-Integrated Systems. New York: Marcel-Dekker Inc.

Kumaran, T. K., W. Chang, H. Cho, and A. Wysk. 1994. A structured approach to deadlock detection, avoidance and resolution in flexible manufacturing systems. International Journal of Production Research. 32(10): 2361–2379.

Li, Z. W., H. S. Hu, and A. R. Wang. 2007. Design of liveness-enforcing supervisors for flexible manufacturing systems using Petri nets. IEEE Transactions on Systems, Man, and Cybernetics, Part C. 37(4): 517–526.

Li, Z. W. and M. C. Zhou. 2004. Elementary siphons of Petri nets and their application to deadlock prevention in flexible manufacturing systems. IEEE Transactions on Systems, Man, and Cybernetics, Part A. 34(1): 38–51.

Li, Z. W. and M. C. Zhou. 2006. Two-stage method for synthesizing liveness-enforcing supervisors for flexible manufacturing systems using Petri nets. IEEE Transactions on Industrial Informatics. 2(4): 313–325.

Li, Z. W. and M. C. Zhou. 2008. Control of elementary and dependent siphons in Petri nets and their application. IEEE Transactions on Systems, Man, and Cybernetics, Part A. 38(1): 133–148.

Li, Z. W., M. C. Zhou, and M. D. Jeng. 2008. A maximally permissive deadlock prevention policy for FMS based on Petri net siphon control and the theory of regions. IEEE Transactions on Automation Science and Engineering. 5(1): 182–188.

Lind-Nielsen, J. 2002. BuDDy: Binary Decision Diagram Package Release 2.2. IT-University of Copenhagen (ITU).

Minato, S. 2000. Zero-suppressed BDDs and their applications. International Journal on Software Tools for Technology Transfer. 3(2): 156–170.

Miner, A. S. and G. Ciardo. 1999. Efficient reachability set generation and storage using decision diagrams. Lecture Notes in Computer Science. 1639: 6–25.

Pastor, E., O. Roig, J. Cortadella, and R. M. Badia. 1994. Petri net analysis using Boolean manipulation. Lecture Notes in Computer Science. 815: 416–435.

Pastor, E., J. Cortadella, and O. Roig. 2001. Symbolic analysis of bounded Petri nets. IEEE Transactions on Computers. 50(5): 432–448.

Piroddi, L., R. Cordone, and I. Fumagalli. 2008. Selective siphon control for deadlock prevention in Petri nets. IEEE Transactions on Systems, Man, and Cybernetics, Part A. 38(6): 1337–1348.

Starke, P. H. 2003. INA: Integrated Net Analyzer. http://www2.informatik.huberlin.de /starke/ina.html.

Tricas, F., F. Garcia-Valles, J. M. Colom, and J. Ezpelata. An iterative method for deadlock prevention in FMS. pp. 139–148. In G. Stremersch [ed.]. 2000. Discrete Event Systems: Analysis and Control, Kluwer Academic, Boston MA.

Uzam, M. 2002. An optimal deadlock prevention policy for flexible manufacturing systems using Petri net models with resources and the theory of regions. International Journal of Advanced Manufacturing Technology. 19(3): 192–208.

Uzam, M. and M. C. Zhou. 2006. An improved iterative synthesis method for liveness enforcing supervisors of flexible manufacturing systems. International Journal of Production Research. 44(10): 1987–2030.

Wang, A. R., Z. W. Li, J. Y. Jia, and M. C. Zhou. 2009. An effective algorithm to find elementary siphons in a class of Petri nets. IEEE Transactions on Systems, Man, and Cybernetics, Part A. 39(4): 912–923.

Wu, N. Q. 1999. Necessary and sufficient conditions for deadlock-free operation in flexible manufacturing systems using a colored Petri net model. IEEE Transactions on Systems, Man, and Cybernetics, Part C. 29(2): 192–204.

Wu, N. Q. and M. C. Zhou. 2001. Avoiding deadlock and reducing starvation and blocking in automated manufacturing systems. IEEE Transactions on Robotics and Automation. 17(5): 658–669.

Xing, K. Y., B. S. Hu, and H. X. Chen. 1996. Deadlock avoidance policy for Petri-net modeling of flexible manufacturing systems with shared resources. IEEE Transactions on Automatic Control. 41(2): 289–295.

Yamalidou, K. J., Moody, M. Lemmon, and P. Antsaklis. 1996. Feedback control of Petri nets based on place invariants. Automatica. 32(1): 15–28.

Zhou, M. C. and F. DiCesare. 1991. Parallel and sequential mutual exclusions for Petri net modeling of manufacturing systems with shared resources. IEEE Transactions on Robotics and Automation. 7(4): 515–527.

Chapter 4

Supervisor Design Based on the Theory of Regions

ABSTRACT

This chapter reports the application of the theory of regions in liveness-enforcing supervisor design, which is an effective method that can definitely obtain a maximally permissive Petri net supervisor if such a supervisor exists. It first finds the marking/transition separation instances (MTSIs) through the analysis of the reachability graph of a plant net model. An MTSI is a pair of a marking and a transition. Once the transition of an MTSI fires at the paired marking, the system enters a zone in the reachability graph, from which the initial marking is unreachable. The theory of regions obtains a liveness-enforcing supervisor by computing a control place to prevent the transition from firing at the marking for each MTSI via solving a linear programming problem (LPP).

4.1 INTRODUCTION

Behavioral permissiveness is an important criterion to evaluate the performance of a Petri net supervisor. A behaviorally optimal Petri net supervisor implies that the supervisor does not prohibit any safe state but forbid all unsafe states. In this case, the controlled system can utilize the resources at anytime if the utilization does not lead to an unsafe state. A behaviorally optimal supervisor usually implies high utilization of the resources in AMSs. Thus, it is interesting and significant to achieve the goal of optimal control (Ghaffari et al., 2003; Piroddi et al., 2008, 2009). However, most of the existing deadlock prevention policies are suboptimal (Fanti and Zhou, 2004, 2005; Huang et al., 2001; Jeng and Xie, 2005; Li and Zhou, 2004, 2006; Uzam and Zhou, 2006; Viswanadham et al., 1990). Usually, a behaviorally optimal supervisor is called a maximally permissive supervisor.

The theory of regions (Badouel and Darondeau, 1998) is an effective approach to synthesize Petri nets from finite transition systems. It has been widely used to design Petri net supervisors (Li et al., 2008; Reveliotis and Choi, 2006; Uzam, 2004). Uzam (2002) first provides a method to design optimal Petri net supervisors by using the theory of regions. However, the approach is not easy to use since its computation is very complex. Ghaffari et al. (2003) develop an intuitive approach to design optimal supervisors by using the theory of regions. The approach uses the concepts of Petri nets and linear algebra, and computes control places by solving LPPs. It is easy to use and more importantly, an optimal supervisor of a Petri net model can be definitely found if such a supervisor exists. Both methods proposed by Uzam (2002) and Ghaffari *et al.* (2003) are general, i.e., they can be applied to all classes of Petri nets. In this chapter, the theory of regions and its application to design optimal Petri net supervisors are recalled in detail.

4.2 THE THEORY OF REGIONS

The theory of regions is general since it can find maximally permissive supervisors for all classes of Petri nets (Ghaffari *et al.*, 2003). It is based on reachability graph analysis of a Petri net and proceeds with an iterative way such that a control place is obtained by solving an LPP at each iteration. In the following, we report the main idea of this method.

Let $G(N, M_0)$ be the reachability graph of a Petri net. For deadlock control purposes, markings in the reachability graph of a Petri net are generally classified into four categories: deadlock, bad, dangerous, and good ones. Both good and dangerous markings represent the legal behavior of a system, which, therefore, are called legal states whose set is denoted by M_L. A maximally permissive supervisor should keep the maximum legal behavior of a system.

As stated in Section 2.2.3, a reahability graph can be classified into two parts: a live-zone (LZ) and a deadlock-zone (DZ), where the LZ contains all the legal markings and the DZ includes all deadlock and bad markings that inevitably lead to deadlock states. A marking/transition separation instance (MTSI) is a pair of a reachable marking M and a transition t that is enabled at M, which is denoted as (M, t). If $M[t\rangle M'$ is true, then M' is in the DZ. Thus, once t fires at M, the system enters the DZ. In this case, M is called a dangerous marking. The set Ω of MTSIs is defined below:

$$\Omega = \{(M, t)|M[t\rangle M' \wedge M \in LZ \wedge M' \in DZ\},$$

where M is a dangerous marking and M' is a bad or deadlock state.

In order to solve the control problem, one has to identify the set of MTSIs in a net from a legal marking to a bad or dead one. An MTSI is a pair of a marking and a transition. Once the transition of an MTSI fires at the paired marking, the system enters a zone in the reachability graph, i.e., DZ, from which the initial marking is unreachable. An additional monitor is used to prohibit the transition from occurring in order to prevent the the system from entering the DZ. Let M_D be the set of dangerous markings. Clearly, we have $M_D = \{M|M \in M_L \wedge \exists t \in T, \ni M[t\rangle M' \wedge M' \notin M_L\}$.

As stated previously, an MTSI is with the form of (M, t), where M is a dangerous marking, and t is a transition whose firing leads to a bad or deadlock marking. Fig. 4.1 shows a Petri net and its reachability graph is shown in Fig. 4.2. In the net, we have $\Omega = \{(M_1, t_5), (M_2, t_1), (M_3, t_5), (M_5, t_1), (M_6, t_5), (M_{11}, t_1)\}$.

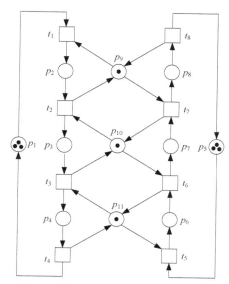

Figure 4.1 A Petri net system (N, M_0).

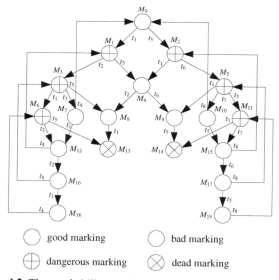

Figure 4.2 The reachability graph of the Petri net (N, M_0) in Fig. 4.1.

Once all MTSIs of a Petri net model are found, we can design a supervisor to separate the DZ from the LZ by prohibiting the transition t from firing at its corresponding marking M for each MTSI (M, t). Suppose that there is a supervisor expressed by a set of control places $\{p_{c_1}, p_{c_2}, \dots, p_{c_{n_c}}\}$ to control the model, whose incidence matrix is of $n_c \times m$, denoted as $[N_c]$, where we use $[N_c](p_c, \cdot)$ to represent the incidence vector of control place p_c and n_c indicates the number of control places. For an expedient description, in the following, $[N_{p_c}]$ denotes $[N_c](p_c, \cdot)$. Suppose that p_c is the monitor to implement an MTSI (M, t). In the following, we provide an ILPP to compute its incidence vector $[N_{p_c}]$ and the initial marking $M_0(p_c)$.

For optimal control purposes, every marking M_l in $\mathcal{M}_L$ must be reachable after the addition of p_c, which implies that p_c has to satisfy the reachability condition, i.e.,

$$M(p_c) = M_0(p_c) + [N_{p_c}]\overrightarrow{\Gamma}_{M_l} \geq 0, \; \forall M_l \in \mathcal{M}_L \tag{4.1}$$

where Γ_{M_l} is any non-oriented path in the LZ from M_0 to M_l, $\overrightarrow{\Gamma}_{M_l}$ is a T-vector, and $\overrightarrow{\Gamma}_{M_l}(t)$ indicates the algebraic sum of all occurrences of t in Γ_{M_l}. $\overrightarrow{\Gamma}$ is called the counting vector of Γ. Similarly, each monitor p_c should satisfy the cycle equation for each cycle in the LZ, i.e.,

$$[N_{p_c}] \cdot \overrightarrow{\gamma} = 0, \forall \gamma \in C_L \tag{4.2}$$

where γ is any non-oriented cycle of the LZ, $\overrightarrow{\gamma}$ is an $m \times 1$ T-vector, $\overrightarrow{\gamma}(t)$ denotes the algebraic sum of all occurrences of t in γ, and C_L is the set of non-oriented cycles in the LZ.

For (M, t), p_c should prohibit transition t from firing at M. Hence, each additional monitor p_c must solve at least one MTSI (M, t) in Ω, i.e.,

$$M_0(p_c) + [N_{p_c}]\overrightarrow{\Gamma}_M + [N_{p_c}](t) \leq -1 \tag{4.3}$$

Eq. (4.3) is called the marking/transition separation equation.

For an MTSI (M, t), Eqs.(4.1), (4.2), and (4.3) are combined as an ILPP, namely the Optimal Supervisor Design by the Theory of Regions (OSDTR), as presented below:

OSDTR(M, t):

min 0

subject to

$$M_0(p_c) + [N_{p_c}]\overrightarrow{\Gamma}_{M_l} \geq 0, \; \forall M_l \in \mathcal{M}_L \tag{4.4}$$

$$[N_{p_c}] \cdot \overrightarrow{\gamma} = 0, \forall \gamma \in C_L \tag{4.5}$$

$$M_0(p_c) + [N_{p_c}]\overrightarrow{\Gamma}_M + [N_{p_c}](t) \leq -1 \tag{4.6}$$

$$M_0(p_c) \in \{0, 1, 2, \ldots\}$$

$$[N_{p_c}](t_i) \in \{0, 1, 2, \ldots\}, \; \forall t_i \in T$$

OSDTR(M, t) can determine the additional monitor p_c, i.e., $[N_{p_c}]$ and $M_0(p_c)$. Then, the obtained control place p_c does not forbid any legal marking and can prevent the firing of t at M. That is to say, p_c is optimal. Note that there is no objective function for OSDTR(M,t) since any feasible solution can lead to an optimal control place p_c to implement (M,t).

We discuss in detail on OSDTR(M,t) with respect to the numbers of constraints and variables. Let us consider the number of constraints of each type. The type of Eq. (4.4) has $|\mathcal{M}_L|$ constraints. The number of the constraints in Eq. (4.5) is $|C_L|$. By considering Eq. (4.6), the total number of all the four types of constraints in OSDTR(M,t) is $|\mathcal{M}_L| + |C_L| + 1$. Now we consider the number of variables in OSDTR(M,t). The number of variables $[N_{p_c}](t_i)$ $(\forall t_i \in T)$ is $|T|$. By considering variable $M_0(p_c)$, OSDTR(M,t) has $|T| + 1$ variables in total. Table 4.1 summarizes the information of OSDTR(M,t).

Table 4.1 The numbers of constraints and variables in OSDTR(M,t)

Constraint type	No. of constraints	Variable name	No. of variables
(4.4)	$|\mathcal{M}_L|$	$[N_{p_c}](t_i)$	$|T|$
(4.5)	$|C_L|$	$M_0(p_c)$	1
(4.6)	1		
total	$|\mathcal{M}_L| + |C_L| + 1$	total	$|T| + 1$

From Table 4.1, we can see that the number of the constraints in OSDTR(M,t) is more than that of legal markings. This is the major bottleneck of the theory of regions and makes it inapplicable to large-scale Petri net models.

In order to make the controlled system running in $\mathcal{M}_L$, each element in Ω should be solved by the addition of a monitor p_c. That is to say, for each MTSI $(M,t) \in \Omega$, the firing of t at M must be prohibited by at least one control place. An iterative approach, formalized as *Algorithm* 1, is developed in (Ghaffari *et al.*, 2003) to compute monitors added to the plant net model, which can lead to a liveness-enforcing net supervisor (LENS). First, the reachability graph is computed and the set Ω of MTSIs is identified. At each iteration, an MTSI is singled out and an optimal control place is designed to implement the MTSI by solving an ILPP. Then, the computed control place is added to the system. Remove all the MTSIs from Ω which are implemented by the control place. The process continues until Ω is empty, which implies that the controlled system is live. Finally, a set of control places are obtained, i.e., the optimal supervisor. The algorithm needs to compute the reachability graph only once and the liveness of the Petri net model is checked by verifying whether all MTSIs are implemented by the computed control places.

A control place may implement more than one MTSI. Let Ω_{p_c} be the set of MTSIs that are implemented by p_c. Thus, we have

Definition 4.1 $\Omega_{p_c} = \{(M,t) \in \Omega | M_0(p_c) + [N_{p_c}]\vec{I}_M + [N_{p_c}](t) \leq -1\}$

Corollary 4.1 *All MTSIs in Ω_{p_c} can be implemented by p_c.*

Proof The proof is trivial. □

In this case, the following algorithm can definitely find an optimal Petri net supervisor on the premise that such a supervisor exists.

Algorithm 4.1 (Ghaffari *et al.*, 2003) The Theory of Regions

Input: A Petri net model (N, M_0).
Output: An optimally controlled Petri net system (N^α, M_0^α) if it exists.

1) Compute the reachability graph of (N, M_0).
2) Compute the set of legal markings M_L for (N, M_0).
3) Compute the set of MTSIs Ω for (N, M_0).
4) $V_M := \emptyset$. /* V_M is used to denote the set of control places to be computed.*/
5) **while** $\{\Omega \neq \emptyset\}$ **do**

$\forall (M, t) \in \Omega$, design OSDTR$(M, t)$ presented in this section.

if $\{$OSDTR(M, t) has no solution$\}$ **then**

Exit, as (N, M_0) has no optimal Petri net supervisor.

else

Compute a control place p_c to implement (M, t).
$V_M := V_M \cup \{p_c\}$ and $\Omega := \Omega - \Omega_{p_c}$.
Denote the resulting net system as (N^α, M_0^α).

endif
endwhile

6) Output (N^α, M_0^α).
7) End.

The next result from (Ghaffari *et al.*, 2003) indicates the existence of an optimal liveness-enforcing supervisor.

Theorem 4.1 *There exists an optimal liveness-enforcing supervisor for a plant Petri net model (N, M_0) iff there exists a set of monitors that implement all MTSIs of (N, M_0).*

Corollary 4.2 $\{M|_P | M \in R((N^\alpha, M_0^\alpha)\} = M_L$ *if (N^α, M_0^α) is an optimal controlled system for (N, M_0), where $M|_P$ denotes the confined making of M to place set P that is the set of places in N.*

The approaches for the design of optimal liveness-enforcing supervisors in (Ghaffari *et al.*, 2003) and (Uzam, 2002) need the complete state enumeration of a plant net model, whose computation is expensive or impossible when we deal with either a large-sized net model, or a small-sized one with a large initial marking. The work of (Ghaffari *et al.*, 2003) shows that, in general, the existence of an optimal liveness-enforcing supervisor can be determined by first generating the reachability graph, and then solving an LPP for each MTSI. Unfortunately, the number of LPPs that need to solve is in theory exponential with respect to the size of the plant net and the initial marking. Furthermore, the number of constraints in such an LPP is in theory exponential with respect to the net size and the initial marking. Although an LPP can be solved in polynomial time, the complexity of the supervisor synthesis approach based on the theory of regions makes it actually impractical.

Though the theory of regions has the disadvantages of complex computation and too many additional monitors, it is still an effective method to find an optimal Petri net supervisor. First, it is general since it can be applied for all classes of Petri nets. Second, it can definitely find an optimal supervisor if it exists. Thus, many researchers use it to find optimal supervisors and try to improve its efficiency.

4.3 AN ILLUSTRATIVE EXAMPLE

In this section, the Petri net model shown in Fig. 4.1 is considered to illustrate the above procedure. In this net, there are six MTSIs, i.e., $\Omega = \{(M_1, t_5), (M_2, t_1), (M_3, t_5), (M_5, t_1), (M_6, t_5), (M_{11}, t_1)\}$. First (M_1, t_5) is considered. A monitor p_{c_1} is designed to implement this MTSI. For simplicity, let $x = M_0^\alpha(p_{c_1})$ and $x_i = [N^\alpha](p_{c_1}, t_i)$, where $i \in \{1, 2, \ldots, 8\}$. M_L has totally 15 markings, i.e., $M_L = \{M_0 - M_3, M_5 - M_7, M_{10} - M_{12}, M_{15} - M_{19}\}$. For optimal control purposes, p_{c_1} should satisfy the reachability conditions, as presented below.

$$x \geq 0 \tag{4.7}$$

$$x + x_1 \geq 0 \tag{4.8}$$

$$x + x_1 + x_2 \geq 0 \tag{4.9}$$

$$x + x_1 + x_2 + x_1 \geq 0 \tag{4.10}$$

$$x + x_1 + x_2 + x_3 \geq 0 \tag{4.11}$$

$$x + x_1 + x_2 + x_3 + x_1 \geq 0 \tag{4.12}$$

$$x + x_1 + x_2 + x_3 + x_1 + x_2 \geq 0 \tag{4.13}$$

$$x + x_1 + x_2 + x_3 + x_1 + x_2 + x_1 \geq 0 \tag{4.14}$$

$$x + x_5 \geq 0 \tag{4.15}$$

$$x + x_5 + x_6 \geq 0 \tag{4.16}$$

$$x + x_5 + x_6 + x_5 \geq 0 \tag{4.17}$$

$$x + x_5 + x_6 + x_7 \geq 0 \tag{4.18}$$

$$x + x_5 + x_6 + x_7 + x_5 \geq 0 \tag{4.19}$$

$$x + x_5 + x_6 + x_7 + x_5 + x_6 \geq 0 \tag{4.20}$$

$$x + x_5 + x_6 + x_7 + x_5 + x_6 + x_5 \geq 0 \tag{4.21}$$

Next, we consider the circuit equations. Note that different circuits in Fig. 4.2 may have the same circuit equations. For example, $M_0 t_1 M_1 t_2 M_3 t_3 M_7 t_4 M_0$ and $M_{12} t_2 M_{16} t_1 M_{18} t_4 M_6 t_3 M_{12}$ are two circuits. It is easy to verify that they have the same circuit equations. Therefore, we have:

$$x_1 + x_2 + x_3 + x_4 = 0 \tag{4.22}$$

$$x_5 + x_6 + x_7 + x_8 = 0 \tag{4.23}$$

Finally, we have the following marking/transition separation equation:

$$x + x_1 + x_5 \leq -1 \qquad (4.24)$$

Solving inequalities (4.7)–(4.24), we can obtain a solution with $x = 1$, $x_1 = -1, x_2 = 1, x_5 = -1, x_6 = 1$, and $x_3 = x_4 = x_7 = x_8 = 0$. Hence monitor p_{c_1} is accordingly added, as shown in Fig. 4.3, to the original Petri net, where $M_0(p_{c_1}) = 1$, $^\bullet p_{c_1} = \{t_2, t_6\}$, and $p_{c_1}^\bullet = \{t_1, t_5\}$.

Now we consider MTSI (M_2, t_1). It is obvious to see that the reachability conditions and circuit equations for all MTSIs are the same, i.e., Eqs. (4.7) – (4.23). When we deal with a new MTSI, we need to replace Eq. (4.24) with the corresponding separation equation only. The separation equation for (M_2, t_1) happens to be Eq. (4.24). Hence we can say that p_{c_1} has implemented (M_2, t_1) as well. Similarly, we can find monitor p_{c_2} that implements (M_3, t_5) and (M_6, t_5) and monitor p_{c_3} that implements (M_5, t_1) and (M_{11}, t_1). Consequently, the resulting controlled system with three monitors is live, as shown in Fig. 4.3.

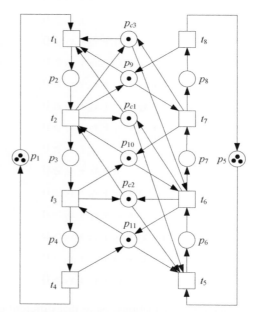

Figure 4.3 A controlled system (N^α, M_0^α).

Note that although the number of monitors to add is theoretically at most equal to, practically much smaller than, the number of MTSIs in the reachability graph of a Petri net, the number of the sets of inequalities that we have to solve is actually equal to that of MTSIs. This is so since we do not know whether a monitor can implement two or more MTSIs until all the sets of inequalities are already solved.

For example, there are 20 reachable markings in Fig. 4.2, where two of them are deadlock markings and six are dangerous markings. Hence there are six MTSIs. We know that three monitors actually implement all the six MTSIs only after we have solved all the six sets of inequalities. This is one of the major disadvantages of the method.

A controlled system is said to be optimal if it results from the synchronous synthesis of a plant net model and an optimal supervisor. For example, the controlled net system shown in Fig. 4.3 is optimal since the supervisor with three monitors is optimal.

4.4 CONCLUSIONS

This chapter provides the theory of regions to design optimal liveness-enforcing Petri net supervisors. The general advantage of the approach is that it can definitely find an optimal Petri net supervisor if such a supervisor exists. However, it suffers from computational and structural complexity problems.

The presented method suffers from the computational complexity problem since a reachability graph is required. In order to enhance the efficiency, one can use BDDs (Andersen, 1997; Brant, 1992; Pastor *et al.*, 1994, 2001) to compute a reachability graph, as stated in Chapter 3.

In order to reduce the computational overheads of the theory of regions, a hybrid method, namely the two-stage method, is proposed in (Li *et al.*, 2008) to improve the efficiency of the theory of regions. The two-stage method includes a siphon solution approach in S^3PR and synthesis of a net supervisor while lowering the computational costs when using the theory of regions. The deadlock prevention policy in (Ezpeleta *et al.*, 1995) is of exponential complexity since a complete siphon enumeration is needed. Moreover, the supervisors obtained in (Ezpeleta *et al.*, 1995) are not maximally permissive in general. The combined approach in (Li *et al.*, 2008) can provide the maximally permissive LENS for an S^3PR model and is more efficient than using the theory of regions alone.

4.5 BIBLIOGRAPHICAL REMARKS

Most materials in this chapter can be found in (Ghaffari *et al.*, 2003). Li *et al.* (2008) provide a two-stage deadlock control policy to improve the computational efficiency. The theory of regions can be found in (Badouel and Darondeau, 1998), which is first used to synthesize Petri net supervisors in (Uzam, 2002). Then Ghaffari *et al.* (2003) propose a plain and popular linear algebraic notion to design optimal Petri net supervisors. Readers can also refer to (Huang *et al.*, 2012) for a new improved version of the theory of regions.

References

Andersen, H. R. 1997. An introduction to binary decision diagrams. Lecture Notes for 49285 Advanced Algorithms E97, Department of Information Technology, Technical University of Denmark.

Badouel, E. and P. Darondeau. Theory of regions. pp. 529–586. *In* W. Reisig and G. Rozenberg, [eds.]. 1998. Lectures on Petri Nets I: Basic Models, Lecture Notes in Compute Science. Berlin: Springer-Verlag, Germany.

Barkaoui, K. and B. Lemaire. 1989. An effective characterization of minimal deadlocks and traps in Petri nets based on graph theory, 1–21. *In* Procedings of the 10th International conference on Application and Theory of Petri Nets, Bonn, Germany.

Brant, R. 1992. Symbolic Boolean manipulation with ordered binary decision diagrams. ACM Computing Surveys. 24(3): 293–318.

Coffman, E. G., M. J. Elphick, and A. Shoshani. 1971. System deadlocks. ACM Computing Surveys. 3(2): 67–78.

Cormen, T. H., C. E. Leiserson, and R. L. Rivest. 1992. Introduction to Algorithms. The MIT Press/MacGraw–Hill Book Company.

Ezpeleta, J., J. M. Colom, and J. Martinez. 1995. A Petri net based deadlock prevention policy for flexible manufacturing systems. IEEE Transactions on Robotics and Automation. 11(2): 173–184.

Fanti, M. P. and M. C. Zhou. 2004. Deadlock control methods in automated manufacturing systems. IEEE Transactions on Systems, Man, and Cybernetics, Part A. 34(1): 5–22.

Fanti, M. P. and M. C. Zhou. Deadlock control methods in automated manufacturing systems. pp. 1–22. *In* M. C. Zhou and M. P. Fanti. [eds.]. 2005. Deadlock Resolution in Computer-Integrated Systems. New York: Marcel-Dekker Co.

Ghaffari, A., N. Rezg, and X. L. Xie. 2003. Design of a live and maximally permissive Petri net controller using the theory of regions. IEEE Transactions on Robotics and Automation. 19(1): 137–142.

Huang, Y. S., M. D. Jeng, X. L. Xie, and S. L. Chung. 2001. Deadlock prevention based on Petri nets and siphons. International Journal of Production Research. 39(2): 283–305.

Huang, Y. S., Y. L. Pan, and M. C. Zhou. 2012. Computationally improved optimal deadlock control policy for flexible manufacturing systems. IEEE Transactions Systems, Man, and Cybernetics, Part A. 42(2): 404–415.

Jeng, M. D. and X. L. Xie. Deadlock detection and prevention of automated manufacturing systems using Petri nets and siphons. pp. 233–281. *In* M. C. Zhou and M. P. Fanti. [eds.]. 2005. Deadlock Resolution in Computer-Integrated Systems. New York: Marcel-Dekker Inc.

Lautenbach, K. and H. Ridder. 1994. Liveness in bounded Petri nets which are covered by T-invariants. Lecture Notes in Computer Science. 815: 358–375.

Li, Z. W. and M. C. Zhou. 2004. Elementary siphons of Petri nets and their application to deadlock prevention in flexible manufacturing systems. IEEE Transactions on Systems, Man, and Cybernetics, Part A. 34(1): 38–51.

Li, Z. W. and M. C. Zhou. 2006. Two-stage method for synthesizing liveness-enforcing supervisors for flexible manufacturing systems using Petri nets. IEEE Transactions on Industrial Informatics. 2(4): 313–325.

Li, Z. W., M. C. Zhou, and M. D. Jeng. 2008. A maximally permissive deadlock prevention policy for FMS based on Petri net siphon control and the theory of regions. IEEE Transactions on Automation Science and Engineering. 5(1): 182–188.

Pastor, E., O. Roig, J. Cortadella, and R. M. Badia. 1994. Petri net analysis using Boolean manipulation. Lecture Notes in Computer Science. 815: 416–435.

Pastor, E., J. Cortadella, and O. Roig. 2001. Symbolic analysis of bounded Petri nets. IEEE Transactions on Computers. 50(5): 432–448.

Piroddi, L., R. Cordone, and I. Fumagalli. 2008. Selective siphon control for deadlock prevention in Petri nets. IEEE Transactions on Systems, Man, and Cybernetics, Part A. 38(6): 1337–1348.

Piroddi, L., R. Cordone, and I. Fumagalli. 2009. Combined siphon and marking generation for deadlock prevention in Petri nets. IEEE Transactions on Systems, Man and Cybernetics, Part A. 39(3): 650–661.

Reveliotis, S. A. and J. Y. Choi. 2006. Designing reversibility-enforcing supervisors of polynomial complexity for bounded Petri nets through the theory of regions. Lecture Notes in Computer Science. 4024: 322–341.

Uzam, M. 2002. An optimal deadlock prevention policy for flexible manufacturing systems using Petri net models with resources and the theory of regions. International Journal of Advanced Manufacturing Technology. 19(3): 192–208.

Uzam, M. 2004. The use of the Petri net reduction approach for an optimal deadlock prevention policy for flexible manufacturing systems. International Journal of Advanced Manufacturing Technology. 23(3–4): 204–219.

Uzam, M. and M. C. Zhou. 2006. An improved iterative synthesis method for liveness enforcing supervisors of flexible manufacturing systems. International Journal of Production Research. 44(10): 1987–2030.

Viswanadham, N., Y. Narahari, and T. Johnson. 1990. Deadlock prevention and deadlock avoidance in flexible manufacturing systems using Petri net models. IEEE Transactions on Robotics and Automation. 6(6): 713–723.

Yamalidou, K. J., Moody, M. Lemmon, and P. Antsaklis. 1996. Feedback control of Petri nets based on place invariants. Automatica. 32(1): 15–28.

Chapter 5

Maximally Permissive Supervisors

ABSTRACT

This chapter presents a computationally efficient method to design optimal control places, and an iteration approach that computes the reachability graph only once in order to obtain a maximally permissive liveness-enforcing supervisor for an AMS. By studying the relationship between different markings, we present a vector covering approach that can remarkably reduce the number of the legal markings and first-met bad markings (FBMs) to be considered. Then, a minimal covering set of legal markings and a minimal covered set of FBMs are found. At each iteration, an FBM from the minimal covered set is selected. In this case, an ILPP is designed to compute a place-invariant (PI) that prevents the FBM from being reached but no marking in the minimal covering set of legal markings. This process is carried out until no FBM is reachable. Also, we use BDDs to compute the sets of legal markings and FBMs presented in Chapter 3 and provide an approach to solve the vector covering problem by using BDDs, aiming to find a minimal covering set of legal markings and a minimal covered set of FBMs. The application of BDDs for reachability graph analysis makes the considered problem computationally tractable and applicable to large-scale Petri net models. Finally, a number of AMS examples are presented to illustrate the presented approaches.

5.1 INTRODUCTION

The theory of regions is an effective approach that can derive a maximally permissive supervisor for a Petri net model by adding monitors if such a supervisor exists (Ghaffari *et al.*, 2003). Thus it has been widely used for deadlock prevention (Li *et al.*, 2008; Uzam, 2002). However, its computation is rather inefficient because it requires the reachability graph of a net, which always suffers from the state explosion problem. Furthermore, its monitors are designed by solving a large number of sets of inequalities, which suffers from the computational complexity problem. Uzam and Zhou (2006) develop an iterative approach for the deadlock control in AMSs. In their study, a reachability graph is classified into a deadlock-zone including deadlock and critical bad markings that inevitably lead to deadlocks, and a live-zone representing the legal markings in the reachability graph. An FBM is singled out from the reachability graph at each iteration and then a control place is designed to prevent the FBM from being reached via constructing a PI of a Petri net by using a well-established invariant-based control method to compute a control place for the PI (Lautenbach and Ridder, 1994; Yamalidou *et al.*, 1996). Although the method proposed in (Uzam and Zhou, 2006) is easy to use and intuitive, it cannot guarantee the optimality in general for an AMS that is modeled by Petri nets. Piroddi *et al.* (2008) develop a selective siphon control approach that proceeds with an iterative way. At each iteration, the relations between uncontrolled siphons and critical markings (at which at least one siphon is empty) are identified and a set of siphons is selected by solving a set covering problem. Once the selected siphons are controlled, all the critical markings are accordingly forbidden, which indicates that all the uncontrolled siphons are controlled. The selective siphon technique can provide a structurally small-sized supervisor. Piroddi *et al.* (2009) provide an improved method to avoid a full siphon enumeration and greatly reduce the overall computational time. Although for every example presented by Piroddi *et al.* (2008, 2009), they can find a maximally permissive supervisor, it is a pity that their policy is not proven to be maximally permissive in theory.

For the Petri net model of an AMS, this chapter presents a computationally efficient method to design the optimal control places, and an iterative approach that computes the reachability graph only once to obtain an optimal liveness-enforcing Petri net supervisor. A vector covering technique is developed, aiming to obtain a minimal covering set of legal markings and a minimal covered set of FBMs. At each iteration, an FBM from the minimal covered set is selected. By solving an ILPP, a PI associated with a monitor is designed to prevent the FBM from being reached and no marking in the minimal covering set of legal markings is forbidden. This process is carried out until no FBM can be reached. In order to make the considered problem computationally tractable, the BDD-based methods presented in Chapter 3 are used to compute the sets of legal markings and FBMs, and solve the vector covering problem to compute a minimal covering set of legal markings and a minimal covered set of FBMs. Finally, a number of AMS examples are presented to illustrate the presented approaches.

This method can be used to design a maximally permissive supervisor for all AMS-oriented classes of Petri net models if such a supervisor exists, such as PPN (Banaszak and Krogh, 1990; Hsieh and Chang, 1994; Xing *et al.*, 1995), S^3PR (Ezpeleta *et al.*, 1995), ES^3PR (Tricas *et al.*, 1998), S^4PR (Tricas *et al.*, 2000), S^*PR (Ezpeleta *et al.*, 2002), S^2LSPR (Park and Reveliotis, 2000), S^3PGR^2 (Park and Reveliotis, 2001), and S^3PMR (Huang *et al.*, 2006). Since a complete enumeration of reachable markings is required, the Petri nets under consideration should be bounded. The application scope of this method is the same as Uzam and Zhou's method but smaller than the theory of regions (Ghaffari *et al.*, 2003) since this method is limited to Petri net models of AMSs but the theory of regions has no such a limitation.

5.2 CONTROL PLACE COMPUTATION

This section reviews the method of designing a control place for a PI proposed in (Yamalidou *et al.*, 1996) and develops an integer linear programming model to obtain an optimal PI. Thus, the control place computed for the optimal PI is also optimal from the viewpoint of deadlock prevention.

5.2.1 Control Place Computation for a Place Invariant

Yamalidou *et al.* (1996) develop a computationally efficient method to construct a supervisor based on the concept of PIs. The control purpose is to enforce algebraic constraints containing elements of a marking. Let $[N_0]$ be the incidence matrix of a plant net. The control places can be represented by a matrix $[N_c]$ that contains the arcs connecting the control places and the transitions of the plant. The controlled net with incidence matrix $[N]$ consists of the original net and the control places, i.e.,

$$[N] = \begin{bmatrix} N_0 \\ N_c \end{bmatrix}$$

The control goal is to enforce the plant to satisfy the following constraint:

$$\sum_{i=1}^{n} l_i \cdot \mu_i \leq \beta \tag{5.1}$$

where μ_i denotes the marking of place p_i, and l_i and β are non-negative integer constants. After the introduction of a non-negative slack variable μ_s, Eq. (5.1) becomes

$$\sum_{i=1}^{n} l_i \cdot \mu_i + \mu_s = \beta \tag{5.2}$$

where μ_s represents the marking of control place p_s, sometimes called a monitor. All constraints in the form of Eq. (5.1) can be grouped into a matrix form and rewritten as follows:

$$[L] \cdot \mu_p \leq b \tag{5.3}$$

where μ_p is the marking vector of the Petri net model, $[L]$ is an $n_c \times n$ nonnegative integer matrix, b is an $n_c \times 1$ nonnegative integer vector, and n_c is the number of constraints. By introducing a non-negative slack variable vector μ_c, these inequality constraints can be transformed into equalities:

$$[L] \cdot \mu_p + \mu_c = b \tag{5.4}$$

where μ_c is an $n_c \times 1$ vector that represents the marking of the control places. According to PI equation $I^T[N] = \mathbf{0}^T$, the supervisor $[N_c]$ can be computed as follows:

$$I^T[N] = [L \; I] \cdot \begin{bmatrix} N_0 \\ N_c \end{bmatrix} = \mathbf{0}^T$$

$$\Leftrightarrow [L] \cdot [N_0] + [N_c] = \mathbf{0}^T$$

$$\Leftrightarrow [N_c] = -[L] \cdot [N_0] \tag{5.5}$$

where I is an $n_c \times n_c$ identity matrix.

Eq. (5.4) must be true at the initial marking μ_0 of a net. Thus, the initial marking μ_{c_0} of the supervisor can be calculated as follows:

$$\mu_{c_0} = b - [L] \cdot \mu_0 \tag{5.6}$$

Definition 5.1 Let I be a PI defined by Eq. (5.2) and p_s the control place associated with I. I is said to be maximally permissive if the addition of p_s can prohibit FBMs only but no legal marking.

The control place computed by a maximally permissive PI is also maximally permissive in the sense that its addition excludes the reachability of FBMs but does not forbid any legal marking.

5.2.2 Optimal Control Place Synthesis

Uzam and Zhou (2006) use the method proposed in (Yamalidou *et al.*, 1996) to compute control places for PIs. For a Petri net model of an AMS, the set of places can be classified into three groups: idle, operation (activity), and resource places whose sets are denoted as P^0, P_A, and P_R, respectively (Ezpeleta *et al.*, 1995; Park and Reveliotis, 2001; Tricas and Martinez, 1995; Zhou and DiCesare, 1991). Tokens in an idle place represent the number of concurrent operations that can happen in a production sequence. An operation place represents an operation to be processed for a part in a production sequence and initially it has no token. Resource places represent resources (machines and robots, for example) and their initial tokens represent the number of available resource units. In (Uzam and Zhou, 2006), for an FBM, the markings of operation places are considered only to obtain a PI to prevent the FBM from being reached.

Given a Petri net model (N, M_0) of an AMS and its reachability graph $G(N, M_0)$, i.e., the set of legal markings $\mathcal{M}_L$ and the set of FBMs $\mathcal{M}_{\mathrm{FBM}}$, this section reports an approach to design control places such that all FBMs are forbidden and all legal markings are reachable. Thus, an optimal supervisor can be obtained. Similar to (Uzam and Zhou, 2006), the tokens in operation places are considered only to obtain a PI to prevent an FBM from being reached. In the following, $\mathbb{N}_A$ denotes $\{i | p_i \in P_A\}$.

Consider an FBM $M \in \mathcal{M}_{\mathrm{FBM}}$. To forbid M, the plant is enforced to satisfy the following constraint:

$$\sum_{i \in \mathbb{N}_A} l_i \cdot \mu_i \leq \beta \tag{5.7}$$

where

$$\beta = \sum_{i \in \mathbb{N}_A} l_i \cdot M(p_i) - 1 \tag{5.8}$$

Eq. (5.7) is called the forbidding condition.

In (Uzam and Zhou, 2006), each coefficient l_i ($i \in \mathbb{N}_A$) is set to 1. Thus, there may exist some legal markings that are forbidden if a control place is added to forbid a selected FBM. For maximally permissive control purposes, all legal markings should be kept after a control place is added. In the following, we show a technique to achieve this purpose. To ensure that every marking $M' \in \mathcal{M}_L$ cannot be prevented from being reached, coefficients l_i's ($i \in \mathbb{N}_A$) should satisfy

$$\sum_{i \in \mathbb{N}_A} l_i \cdot M'(p_i) \le \beta, \quad \forall M' \in \mathcal{M}_L \tag{5.9}$$

Eq. (5.9) is called the reachability condition. For an FBM M, substituting the expression of β in Eq. (5.8) into Eq. (5.9), the reachability condition becomes

$$\sum_{i \in \mathbb{N}_A} l_i \cdot (M'(p_i) - M(p_i)) \le -1, \quad \forall M' \in \mathcal{M}_L \tag{5.10}$$

Eq. (5.10) determines the feasible values for coefficients l_i's ($i \in \mathbb{N}_A$). Thus, for an FBM M, if coefficients l_i's ($i \in \mathbb{N}_A$) satisfy Eq. (5.10), a PI designed for Eq. (5.7) can guarantee the reachability of all the legal markings and forbid M. In this case, a PI can be designed by the method presented in Section 5.2.1. Therefore, a control place computed for the PI can ensure all the legal markings being reached. Then, we claim that the control place is optimal.

5.3 VECTOR COVERING APPROACH FOR PLACE INVARIANT CONTROL

This section presents a vector covering approach for the PI control such that:

1. For the set of FBMs, the relation between different elements is used to obtain a minimal covered set of FBMs. For each FBM in the minimal covered set, a PI is designed to prevent it from being reached. Thus all the FBMs cannot be reached and furthermore unnecessary FBMs can be avoided being considered. Similarly, a minimal covering set of legal markings can be obtained. The goal is that when a PI is designed for an FBM, no marking in the minimal covering set is forbidden. Hence, all the legal markings can be reached. This technique can reduce the computational overheads for every PI.
2. For each element in the minimal covered set of FBMs, by solving an ILPP, a PI is designed to forbid the FBM and keep all the markings in the minimal covering set of legal markings. A control place is computed to implement each PI. Thus a liveness-enforcing supervisor can be obtained.
3. The approach is applied to bounded Petri net models of AMSs.

Definition 5.2 Let M and M' be two markings in $R(N, M_0)$. M A-covers M' (or M' is A-covered by M) if $\forall p \in P_A, M(p) \geq M'(p)$, which is denoted as $M \geq_A M'$ (or $M' \leq_A M$).

This definition is used to study the relations among different markings. The tokens in operation places are considered only to design control places. Thus, the relations between different markings are simplified to study the number of tokens in these operation places. Given this definition, we have the following theorem.

Theorem 5.1 Let M and M' be two markings in $R(N, M_0)$ with $M \geq_A M'$. If M' is forbidden by a PI, M is forbidden. If M is not forbidden by a PI, M' is not forbidden.

Proof Suppose that marking M' is forbidden by a PI that satisfies constraint $\sum_{i \in \mathbb{N}_A} l_i \cdot \mu_i \leq \beta$. In this case, we have $\beta = \sum_{i \in \mathbb{N}_A} l_i \cdot M'(p_i) - 1$. Since $M \geq_A M'$, we have $\forall i \in \mathbb{N}_A, M(p_i) \geq M'(p_i)$. It follows that $\sum_{i \in \mathbb{N}_A} l_i \cdot M(p_i) \geq \sum_{i \in \mathbb{N}_A} l_i \cdot M'(p_i) = \beta + 1$. This means that M is forbidden by the PI. Accordingly, we conclude that if M is not forbidden by a PI, M' is not forbidden. $\square$

Theorem 5.1 indicates that if a PI is designed to prevent a marking M' from being reached, then any marking M that satisfies $M \geq_A M'$ is also unreachable. Similarly, if a PI is designed to keep a marking M being reached, then any marking M' that satisfies $M \geq_A M'$ can also be reached. This theorem is essential to reduce the set of FBMs and the set of legal markings to be small.

Definition 5.3 Let $\mathcal{M}_{FBM}$ be the set of FBMs in a Petri net. $\forall M \in \mathcal{M}_{FBM}$, a subset of $\mathcal{M}_{FBM}$ that A-covers M is defined as $F_M = \{M' \in \mathcal{M}_{FBM} | M' \geq_A M\}$.

Corollary 5.1 If an FBM $M \in \mathcal{M}_{FBM}$ is forbidden by a PI, all the markings in F_M are forbidden.

Proof It follows immediately from Theorem 5.1. $\square$

Definition 5.3 and Corollary 5.1 mean that if an FBM M is forbidden by a PI, then, all the markings in set F_M are forbidden. That is to say, if M is prevented from being reached, there is no need for all other markings in F_M to be prevented from being reached.

Definition 5.4 Let $\mathcal{M}_L$ be the set of legal markings. $\forall M \in \mathcal{M}_L$, a subset of $\mathcal{M}_L$ that is A-covered by M is defined as $R_M = \{M' \in \mathcal{M}_L | M' \leq_A M\}$.

Corollary 5.2 If a legal marking $M \in \mathcal{M}_L$ is not forbidden by a PI, no marking in R_M is forbidden.

Proof It follows immediately from Theorem 5.1. $\square$

Definition 5.4 and Corollary 5.2 mean that if a legal marking M is not forbidden by a PI, then, all the markings in set R_M are not forbidden. That is to say, if M can be reached, all markings in R_M can be reached and there is no need to consider their reachability.

Definition 5.5 Let $\mathcal{M}^{\star}_{\text{FBM}}$ be a subset of FBMs. $\mathcal{M}^{\star}_{\text{FBM}}$ is called the minimal covered set of FBMs if the following two conditions are satisfied:

1. $\forall M \in \mathcal{M}_{\text{FBM}}, \exists M' \in \mathcal{M}^{\star}_{\text{FBM}}, \ s.t. \ M \geq_A M'$; and
2. $\forall M \in \mathcal{M}^{\star}_{\text{FBM}}, \nexists M'' \in \mathcal{M}^{\star}_{\text{FBM}}, \ s.t. \ M \geq_A M''$ and $M \neq M''$.

Corollary 5.3 *If all markings of $\mathcal{M}^{\star}_{\text{FBM}}$ are forbidden by PIs, all the FBMs are forbidden.*

Proof It follows immediately from Definition 5.5 and Corollary 5.1.   □

For the deadlock control purposes, Definition 5.5 and Corollary 5.3 indicate that a small set of FBMs is considered only to prevent all FBMs from being reached. If all markings in a minimal covered set of FBMs are prevented from being reached, all FBMs are forbidden. Thus, in a deadlock prevention policy, the minimal covered set of FBMs is considered only.

Definition 5.6 Let $\mathcal{M}^{\star}_L$ be a subset of legal markings. $\mathcal{M}^{\star}_L$ is called the minimal covering set of legal markings if the following two conditions are satisfied:

1. $\forall M \in \mathcal{M}_L, \exists M' \in \mathcal{M}^{\star}_L \ s.t. \ M' \geq_A M$; and
2. $\forall M \in \mathcal{M}^{\star}_L, \nexists M'' \in \mathcal{M}^{\star}_L \ s.t. \ M'' \geq_A M$ and $M \neq M''$.

Corollary 5.4 *If no marking in $\mathcal{M}^{\star}_L$ is forbidden by PIs, no legal marking is forbidden.*

Proof It follows immediately from Definition 5.6 and Corollary 5.2.   □

For the maximally permissive control purposes, Definition 5.6 and Corollary 5.4 indicate that a small set of legal markings can be considered only to keep all legal markings being reached. If all markings in the minimal covering set of legal markings can be reached, all legal markings can also be reached. Thus, in a deadlock prevention policy, the minimal covering set of legal markings is considered only to obtain a maximally permissive supervisor.

Definition 5.7 Let $\mathcal{M}^{\star}_{\text{FBM}}$ be the minimal covered set of FBMs. If a PI I satisfies constraint $\sum_{i \in \mathbb{N}_A} l_i \cdot \mu_i \leq \beta$, then the set of FBMs that are forbidden by the PI is defined as $F_I = \{M \in \mathcal{M}^{\star}_{\text{FBM}} | \sum_{i \in \mathbb{N}_A} l_i \cdot M(p_i) \geq \beta + 1\}$.

Definition 5.7 means that a PI may prevent more than one FBM from being reached. According to Corollaries 5.3 and 5.4, a supervisor is optimal if it forbids all the markings in the minimal covered set of FBMs and does not forbid any marking in the minimal covering set of legal markings. This indicates that the sets $\mathcal{M}^{\star}_{\text{FBM}}$ and $\mathcal{M}^{\star}_L$ are required to be considered only for the design of control places. Therefore, for an FBM M, the reachability condition, i.e., Eq. (5.10), can be reduced as follows:

$$\sum_{i \in \mathbb{N}_A} l_i \cdot (M'(p_i) - M(p_i)) \leq -1, \quad \forall M' \in \mathcal{M}^{\star}_L \tag{5.11}$$

Experimental results show that in general $\mathcal{M}^{\star}_{\text{FBM}}$ and $\mathcal{M}^{\star}_L$ are much smaller than $\mathcal{M}_{\text{FBM}}$ and $\mathcal{M}_L$, respectively. Thus, the computational costs can be greatly reduced.

For an FBM selected from $\mathcal{M}_{\mathrm{FBM}}^{\star}$, it is unnecessary to consider all markings in $\mathcal{M}_L^{\star}$ when designing an optimal control place. In the following, we reduce $\mathcal{M}_L^{\star}$ to be a smaller set for each FBM and the optimality of each control place is still guaranteed.

Theorem 5.2 *For an FBM $M \in \mathcal{M}_{\mathrm{FBM}}^{\star}$, if there exist coefficients l_i's ($i \in \mathbb{N}_A$) that satisfy Eq. (5.11), then a PI designed for Eq. (5.7) with $l_i^{\star}$ instead of l_i can also guarantee the reachability of all markings in $\mathcal{M}_L^{\star}$, where $\forall i \in \mathbb{N}_A$:*

$$l_i^{\star} = \begin{cases} l_i & \text{if } M(p_i) \neq 0 \\ 0 & \text{if } M(p_i) = 0 \end{cases} \tag{5.12}$$

Proof For an FBM $M \in \mathcal{M}_{\mathrm{FBM}}^{\star}$, according to the definition of $l_i^{\star}$ ($i \in \mathbb{N}_A$), $\forall M' \in \mathcal{M}_L^{\star}$, we have

$\sum_{i \in \mathbb{N}_A} l_i^{\star} \cdot (M'(p_i) - M(p_i)) =$

$\sum_{i \in \mathbb{N}_A, M(p_i) \neq 0} l_i^{\star} \cdot (M'(p_i) - M(p_i)) + \sum_{i \in \mathbb{N}_A, M(p_i) = 0} l_i^{\star} \cdot (M'(p_i) - M(p_i)) =$

$\sum_{i \in \mathbb{N}_A, M(p_i) \neq 0} l_i \cdot (M'(p_i) - M(p_i)) \leq$

$\sum_{i \in \mathbb{N}_A, M(p_i) \neq 0} l_i \cdot (M'(p_i) - M(p_i)) + \sum_{i \in \mathbb{N}_A, M(p_i) = 0} l_i \cdot M'(p_i) =$

$\sum_{i \in \mathbb{N}_A, M(p_i) \neq 0} l_i \cdot (M'(p_i) - M(p_i)) + \sum_{i \in \mathbb{N}_A, M(p_i) = 0} l_i \cdot (M'(p_i) - M(p_i)) =$

$\sum_{i \in \mathbb{N}_A} l_i \cdot (M'(p_i) - M(p_i))$

Since $\sum_{i \in \mathbb{N}_A} l_i \cdot (M'(p_i) - M(p_i)) \leq -1$, $\sum_{i \in \mathbb{N}_A} l_i^{\star} \cdot (M'(p_i) - M(p_i)) \leq -1$. This means that coefficients $l_i^{\star}$'s ($i \in \mathbb{N}_A$) can also satisfy the reachability condition, i.e., Eq. (5.11) and lead to a PI that can guarantee the reachability of all markings in $\mathcal{M}_L^{\star}$. □

This theorem indicates that, for each FBM, an optimal control place can be designed by considering only the operation places that have tokens at the FBM. Thus, a further vector covering approach can be developed to reduce the number of markings that need to be considered in $\mathcal{M}_L^{\star}$.

Definition 5.8 For an FBM $M \in \mathcal{M}_{\mathrm{FBM}}^{\star}$, $\forall M_1, M_2 \in \mathcal{M}_L^{\star}$, M_1 M-covers M_2 (or M_2 is M-covered by M_1) if $\forall p \in P_A$ with $M(p) \neq 0$, we have $M_1(p) \geq M_2(p)$, which is denoted as $M_1 \geq_M M_2$ (or $M_2 \leq_M M_1$).

When an FBM is selected to be prevented from being reached, this definition can be used to study the relations among different legal markings. The operation places that have tokens at the FBM are considered only to design control places. Thus, the relations between different legal markings are simplified to study the tokens in these operation places. Given this definition, we have the following theorem.

Theorem 5.3 *Let $M \in \mathcal{M}_{\mathrm{FBM}}^{\star}$ be an FBM and $M_1, M_2 \in \mathcal{M}_L^{\star}$ two legal markings with $M_1 \geq_M M_2$. If M_1 is not forbidden by a PI designed for M, M_2 is not forbidden.*

Proof Suppose that M is forbidden by a PI that satisfies constraint $\sum_{i \in \mathbb{N}_A} l_i \cdot \mu_i \leq \beta$ with $\beta = \sum_{i \in \mathbb{N}_A} l_i \cdot M(p_i) - 1$. According to Theorem 5.2, the constraint can be reduced to be $\sum_{i\in\mathbb{N}_A, M(p_i)\neq 0} l_i \cdot \mu_i \leq \beta$ with $\beta = \sum_{i\in\mathbb{N}_A, M(p_i)\neq 0} l_i \cdot M(p_i) - 1$. In order not to forbid M_1, coefficients l_i's $(i \in \mathbb{N}_A, M(p_i) \neq 0)$ of the PI should satisfy $\sum_{i\in\mathbb{N}_A, M(p_i)\neq 0} l_i \cdot M_1(p_i) \leq \beta$. Since $M_1 \geq_M M_2$, we have $\sum_{i\in\mathbb{N}_A, M(p_i)\neq 0} l_i \cdot M_2(p_i) \leq \sum_{i\in\mathbb{N}_A, M(p_i)\neq 0} l_i \cdot M_1(p_i) \leq \beta$. This means that M_2 is not forbidden by the PI.   □

Theorem 5.3 indicates that if a PI is designed to prevent an FBM $M \in \mathcal{M}_{\text{FBM}}^{\star}$ from being reached and keep a marking $M_1 \in \mathcal{M}_L^{\star}$ being reached, then any marking $M_2 \in \mathcal{M}_L^{\star}$ that satisfies $M_1 \geq_M M_2$ can also be reached. This theorem can be used to reduce the minimal covering set of legal markings into a much smaller one.

Definition 5.9 Let M be a marking in $\mathcal{M}_{\text{FBM}}^{\star}$. $\mathcal{L}_M \subseteq \mathcal{M}_L^{\star}$ is called the minimal covering set of M-related legal markings if the following two conditions are satisfied:

1. $\forall M' \in \mathcal{M}_L^{\star}, \exists M'' \in \mathcal{L}_M$, s.t. $M'' \geq_M M'$; and
2. $\forall M' \in \mathcal{L}_M, \nexists M'' \in \mathcal{L}_M$, s.t. $M'' \geq_M M'$ and $M'' \neq M'$.

Corollary 5.5 *For an FBM $M \in \mathcal{M}_{\text{FBM}}^{\star}$, if M is forbidden by a PI and no marking in the minimal covering set of M-related legal markings is forbidden, then no legal marking is forbidden.*

Proof It follows immediately from Definition 5.9 and Theorem 5.3.   □

According to Corollary 5.5, for an FBM $M \in \mathcal{M}_{\text{FBM}}^{\star}$, if a PI is designed to forbid M but no marking in the minimal covering set of its related legal markings $\mathcal{L}_M$, all the legal markings can be reached if a control place is added for this PI. This indicates that for an FBM M, the minimal covering set of its related legal markings $\mathcal{L}_M$ requires to be considered only to obtain an optimal control place. Therefore, for an FBM $M \in \mathcal{M}_{\text{FBM}}^{\star}$, the reachability condition, i.e., Eq. (5.11), can be reduced as follows:

$$\sum_{i\in\mathbb{N}_A} l_i \cdot (M'(p_i) - M(p_i)) \leq -1, \quad \forall M' \in \mathcal{L}_M \tag{5.13}$$

Generally, $\mathcal{L}_M$ is much smaller than $\mathcal{M}_L^{\star}$. Therefore, the set of the inequalities for each FBM can be reduced to be a much smaller one.

5.4 SYMBOLIC COMPUTATION OF THE VECTOR COVERING APPROACH

Chapter 3 provides a BDD-based method to compute the legal markings and FBMs of a Petri net model. This section considers how to calculate the minimal sets of legal markings and FBMs. For bounded Petri nets, the numbers of tokens in place p at two markings M' and M are represented by two sets of Boolean variables $q^0, q^1, \ldots,$ and q^{k_p} and $p^0, p^1, \ldots,$ and p^{k_p}, respectively. Then the relation $M'(p) \leq M(p)$ can be defined as follows:

$$M'(p) \leq M(p) \equiv (q^{k_p} < p^{k_p})$$
$$+ (q^{k_p} \equiv p^{k_p}) \cdot (q^{k_p-1} < p^{k_p-1})$$

$$+(q^{k_p} \equiv p^{k_p}) \cdot (q^{k_p-1} \equiv p^{k_p-1}) \cdot (q^{k_p-2} < p^{k_p-2})$$

$$\cdots\cdots$$

$$+(q^{k_p} \equiv p^{k_p}) \cdot (q^{k_p-1} \equiv p^{k_p-1})\ldots(p^1 \equiv \omega^1) \cdot (q^0 < p^0)$$
$$+(q^{k_p} \equiv p^{k_p}) \cdot (q^{k_p-1} \equiv p^{k_p-1})\ldots(q^1 \equiv p^1) \cdot (q^0 \equiv p^0)$$

Thus, the relation $M' \leq_A M$ can be defined by the characteristic function $\mathcal{R}_{\leq_A}$:

$$\mathcal{R}_{\leq_A}(M', M) = \prod_{p \in P_A} \left[\sum_{i=0}^{k_p} \left[(q^i < p^i) \cdot \prod_{j=i+1}^{k_p} (q^j \equiv p^j) \right] + \prod_{i=0}^{k_p}(q^i \equiv p^i) \right]$$

The set of markings $\mathcal{M}_{\leq_A L}$ that is A-covered by $\mathcal{M}_L$ with $\mathcal{M}_{\leq_A L} \neq \mathcal{M}_L$ can be defined as:

$$\mathcal{M}_{\leq_A L} = \{M \in \mathcal{M}_L | \forall M \in \mathcal{M}_{\leq_A L}, \exists M' \in \mathcal{M}_L, \text{ s.t. } M \leq_A M' \text{ and } M \neq M'\}$$

Then, it can be calculated by BDDs as follows:

$$\mathcal{M}_{\leq_A L} = \mathcal{R}_{\leq_A}(\mathcal{M}_{\leq_A L}, \mathcal{M}_L) \cdot \prod_{i \in N_A} \prod_{j=0}^{k_{p_i}} q_i^j \oplus p_i^j \cdot \mathcal{M}_L$$

where $\prod_{i \in N_A} \prod_{j=0}^{k_{p_i}} q_i^j \oplus p_i^j$ indicates that $\mathcal{M}_{\leq_A L} \neq \mathcal{M}_L$. The minimal set of legal markings can be computed as follows:

$$\mathcal{M}_L^\star = \mathcal{M}_L - \mathcal{M}_{\leq_A L}$$

Similarly, the minimal set of FBMs can be computed. The relation $M' \geq_A M$ can be defined by the characteristic function $\mathcal{R}_{\geq_A}$:

$$\mathcal{R}_{\geq_A}(M', M) = \prod_{p \in P_A} \left[\sum_{i=0}^{k_p} \left[(q^i > p^i) \cdot \prod_{j=i+1}^{k_p} (q^j \equiv p^j) \right] + \prod_{i=0}^{k_p}(q^i \equiv p^i) \right]$$

The set of markings $\mathcal{M}_{\geq_A \text{FBM}}$ that A-covers $\mathcal{M}_{\text{FBM}}$ with $\mathcal{M}_{\geq_A \text{FBM}} \neq \mathcal{M}_{\text{FBM}}$ can be defined as follows:

$$\mathcal{M}_{\geq_A \text{FBM}} = \{M \in \mathcal{M}_{\text{FBM}} | \forall M \in \mathcal{M}_{\geq_A \text{FBM}}, \exists M' \in \mathcal{M}_{\text{FBM}}, s.t. M \geq_A M' \text{ and } M \neq M'\}$$

Then, it can be calculated by BDDs as follows:

$$\mathcal{M}_{\geq_A \text{FBM}} = \mathcal{R}_{\geq_A}(\mathcal{M}_{\geq_A \text{FBM}}, \mathcal{M}_{\text{FBM}}) \cdot \prod_{i \in \mathbb{N}_A} \prod_{j=0}^{k_{p_i}} q_i^j \oplus p_i^j \cdot \mathcal{M}_{\text{FBM}}$$

Finally, the minimal set of FBMs can be computed as follows:

$$\mathcal{M}_{\text{FBM}}^\star = \mathcal{M}_{\text{FBM}} - \mathcal{M}_{\geq_A \text{FBM}}$$

An algorithm to compute the minimal sets of legal markings and FBMs is presented as follows:

Algorithm 5.1 Computation of the Minimal Sets of Legal Markings and FBMs

Input: A Petri net system (N, M_0).

Output: The minimal sets of legal markings and FBMs, denoted as $\mathcal{M}_L^\star$ and $\mathcal{M}_{\text{FBM}}^\star$, respectively.

1) Compute the set of legal markings $\mathcal{M}_L$ by Algorithm 3.3.
2) Compute the set of FBMs $\mathcal{M}_{\text{FBM}}$ by Algorithm 3.4.
3) $\mathcal{M}_{\leq_A L} := \mathcal{R}_{\leq_A}(\mathcal{M}_{\leq_A L}, \mathcal{M}_L) \cdot \prod_{i \in \mathbb{N}_A} \prod_{j=0}^{k_{p_i}} q_i^j \oplus p_i^j \cdot \mathcal{M}_L$.
4) $\mathcal{M}_L^\star := \mathcal{M}_L - \mathcal{M}_{\leq_A L}$.
5) $\mathcal{M}_{\geq_A \text{FBM}} := \mathcal{R}_{\geq_A}(\mathcal{M}_{\geq_A \text{FBM}}, \mathcal{M}_{\text{FBM}}) \cdot \prod_{i \in \mathbb{N}_A} \prod_{j=0}^{k_{p_i}} q_i^j \oplus p_i^j \cdot \mathcal{M}_{\text{FBM}}$.
6) $\mathcal{M}_{\text{FBM}}^\star := \mathcal{M}_{\text{FBM}} - \mathcal{M}_{\geq_A \text{FBM}}$.
7) Output $\mathcal{M}_L^\star$ and $\mathcal{M}_{\text{FBM}}^\star$.
8) End.

Similar to the computation of $\mathcal{M}_L^\star$, the minimal covering set of FBM-related legal markings $\mathcal{L}_M$ can be calculated. We use two sets of Boolean variables, $q^0, q^1, \ldots$, and q^{k_p} and $p^0, p^1, \ldots$, and p^{k_p}, to represent markings M_2 and M_1, respectively. For an FBM M, the relation $M_2 \leq_M M_1$ can be defined by the characteristic function $\mathcal{R}_{\leq_M}$:

$$\mathcal{R}_{\leq_M}(M_2, M_1) = \prod_{p \in P_A, M(p) \neq 0} \left[\sum_{i=0}^{k_p} \left[(q^i < p^i) \cdot \prod_{j=i+1}^{k_p} (q^j \equiv p^j) \right] + \prod_{i=0}^{k_p} (q^i \equiv p^i) \right]$$

The set of markings $\mathcal{M}_{\leq_M L}$ that are M-covered by $\mathcal{M}_L^\star$ with $\mathcal{M}_{\leq_M L} \neq \mathcal{M}_L^\star$ can be defined as:

$$\mathcal{M}_{\leq_M L} = \{M' \in \mathcal{M}_L^\star | \forall M' \in \mathcal{M}_{\leq_M L}, \exists M'' \in \mathcal{M}_L^\star, s.t.\ M' \leq_A M'' \text{ and } M' \neq M''\}$$

Then, it can be calculated by BDDs as follows:

$$\mathcal{M}_{\leq_M L} = \mathcal{R}_{\leq_M}(\mathcal{M}_{\leq_M L}, \mathcal{M}_L^\star) \cdot \prod_{p_i \in P_A, M(p_i) \neq 0} \prod_{j=0}^{k_{p_i}} q_i^j \oplus p_i^j \cdot \mathcal{M}_L^\star,$$

where $\prod_{p_i \in P_A, M(p_i) \neq 0} \prod_{j=0}^{k_{p_i}} q_i^j \oplus p_i^j$ indicates that $\mathcal{M}_{\leq_M L} \neq \mathcal{M}_L^\star$. The minimal covering set of M-related legal markings can be computed as follows:

$$\mathcal{L}_M = \mathcal{M}_L^\star - \mathcal{M}_{\leq_M L}$$

Finally, a simple algorithm for computing the minimal covering set of FBM-related legal markings is presented as follows.

Algorithm 5.2 Computation of the Minimal Covering Set of FBM-related Legal Markings

Input: The minimal set of legal markings $\mathcal{M}_L^\star$ and an FBM M selected from $\mathcal{M}_{FBM}^\star$.

Output: The minimal covering set of M-related legal markings denoted as $\mathcal{L}_M$.

1) $\mathcal{M}_{\leq ML} := \mathcal{R}_{\leq M}(\mathcal{M}_{\leq ML}, \mathcal{M}_L^\star) \cdot \prod_{p_i \in P_A, M(p_i) \neq 0} \prod_{j=0}^{k_{p_i}} q_i^j \oplus p_i^j \cdot \mathcal{M}_L^\star$.

2) $\mathcal{L}_M := \mathcal{M}_L^\star - \mathcal{M}_{\leq ML}$.

3) Output $\mathcal{L}_M$.

4) End.

5.5 DEADLOCK PREVENTION POLICY

This section presents a deadlock prevention policy whose development is based on the results in the previous sections.

Algorithm 5.3 Optimal Deadlock Prevention Policy

Input: Petri net model (N, M_0) of an AMS.

Output: An optimally controlled Petri net system (N_1, M_1) if it exists.

1) Compute the minimal covered set of FBMs $\mathcal{M}_{FBM}^\star$ and the minimal covering set of legal markings $\mathcal{M}_L^\star$ for (N, M_0) by Algorithm 5.1.

2) $V_M := \emptyset$. /* V_M is used to denote the set of control places to be computed.*/

3) **if** $\{\mathcal{M}_{FBM}^\star = \emptyset\}$ **then**

$\quad$ $N_1 := N, M_1 := M_0$, go to Step 5.

$\quad$ **else**

$\quad$ $\forall M \in \mathcal{M}_{FBM}^\star$, compute the minimal covering set of M-related legal markings $\mathcal{L}_M$ by Algorithm 5.2.

$\quad$ Solve the set of integer linear inequalities of Eq. (5.13). Let l_i $(i \in \mathbb{N}_A, M(p_i) \neq 0)$ be the solution if it exists. Otherwise, exit, as the optimal control place does not exist for (N, M_0).

$\quad$ Design a PI I and a control place p_c by the method presented in Section 5.2.1.

$\quad$ $V_M := V_M \cup \{p_c\}$, $\mathcal{M}_{FBM}^\star := \mathcal{M}_{FBM}^\star - F_I$. Denote the resulting net system as (N, M_0).

$\quad$ **endif**

4) Go to Step 3.

5) Output (N_1, M_1).

6) End.

The above algorithm uses the reachability graph only once. $\mathcal{M}_{FBM}^\star$ and $\mathcal{M}_L^\star$ are required to be considered only for the computation of each control place. This can reduce the number of inequalities for each PI and the total number of FBMs that should be forbidden by PIs. For each FBM M, $\mathcal{M}_L^\star$ is reduced to be smaller, i.e., $\mathcal{L}_M$.

This can further reduce the number of inequalities for each FBM. In the following, some examples are used to illustrate this policy.

Theorem 5.4 *Algorithm 5.3 can obtain a maximally permissive supervisor for a Petri net model of an AMS if there exists a solution that satisfies Eq. (5.13) for each FBM in $M^\star_{\text{FBM}}$.*

Proof Let (N_1, M_1) be the controlled net system that results from adding control places in V_M to (N, M_0). Each place in V_M is designed by a PI for constraint $\sum_{i \in N_A} l_i \cdot \mu_i \leq \beta$, where l_i's $(i \in \mathbb{N}_A)$ satisfy Eq. (5.13). As Section 5.3 describes, all legal markings can be reached and at least one FBM is forbidden by this control place. According to Corollaries 5.3 and 5.5, all the control places in V_M can forbid all the FBMs and all legal markings can be reached. That is to say, a maximally permissive supervisor is obtained.   □

Next, existing results on Petri nets from (Murata, 1989) are introduced. They include the concepts of conservativeness and structural liveness, and some important conclusions.

Definition 5.10 (Murata, 1989) A Petri net N is said to be structurally live if there exists an initial marking M_0 such that (N, M_0) is live.

Definition 5.11 (Murata, 1989) A Petri net N is said to be structurally bounded if it is bounded at any finite initial marking M_0.

Definition 5.12 (Murata, 1989) A Petri net N is said to be conservative if there exists a positive integer $I(p)$ for every place p such that the weighted sum of tokens, $M^T \cdot I = M_0^T \cdot I = \alpha$, for every $M \in R(N, M_0)$ and for any fixed initial marking M_0, where α is a constant.

Theorem 5.5 (Murata, 1989) *A Petri net N is conservative iff there exists a P-vector I of positive integers such that $I^T[N] = \mathbf{0}^T$.*

Property 5.1 (Murata, 1989) If a Petri net is structurally bounded and structurally live, then it is both conservative and consistent.

Theorem 5.5 and Property 5.1 indicate that conservativeness is a necessary condition for structural boundedness and liveness of a Petri net. In other words, if a net is not conservative, it is neither structurally bounded nor structurally live. This result is used to prove the following theorem. In a Petri net model considered in this work, we have:

Assumption 1: Any idle place p_{id} is associated with a minimal P-semiflow $I_{p_{id}}$ such that $\forall p \in \|I_{p_{id}}\| \setminus \{p_{id}\}, p \in P_A$.

Assumption 2: Any resource place r is associated with a minimal P-semiflow I_r such that $\forall p \in \|I_r\| \setminus \{r\}, p \in P_A$, where P_A is the set of operation places in the Petri net model.

The above assumptions are true in the manufacturing-oriented Petri net models in the literature.

Theorem 5.6 *Under Assumptions 1 and 2, the presented deadlock prevention method can lead to a maximal permissive liveness-enforcing supervisor if such a supervisor exists.*

Proof First, we claim that an optimal control place C is necessarily associated with a PI I_C such that $\forall p \in \|I_C\| \setminus \{C\}$, $p \in P_A$. According to Theorem 5.5 and Property 5.1, if a control place is not associated with a PI, then the controlled net is neither structurally bounded nor structurally live. Thus, a control place C necessarily corresponds to a PI I'_C. Without loss of generality, we assume that $\|I'_C\|$ contains operation, resource, and idle places. Let $I_C = I'_C - \sum_{r \in \|I'_C\| \cap P_R} I_r - \sum_{p_{id} \in \|I'_C\| \cap P^0} I_{p_{id}}$. Then I_C is also a PI since any linear combination of PIs of a net is still a PI. According to Assumptions 1 and 2, there exists a PI that involves a control place and operation places only.

Now, we show that if an optimal control place exists, it can be obtained by the presented method. Assume that a maximally permissive supervisor exists for a net model. Then each control place of the supervisor can construct a PI that does not forbid any legal markings and at least one FBM is forbidden. By contradiction, suppose that there exists a PI that does not satisfy Eq. (5.13) for some markings in $\mathcal{L}_M$. As Section 5.3 describes, these markings are forbidden by the PI. This means that the control place designed for this PI is not optimal. Thus, it is proved that any PI for the optimal control purposes satisfies Eq. (5.13). This means that, for each FBM, there exists at least a solution that satisfies Eq. (5.13). According to Theorem 5.4, the presented method can obtain a maximally permissive liveness-enforcing supervisor if such a supervisor exists. □

In the following, two compact multiset formalisms are used to describe markings in $M_L^\star$, $M_{FBM}^\star$, and $\mathcal{L}_M$. Operation places are considered for the markings only in $M_L^\star$ and $M_{FBM}^\star$. Thus, $\sum_{i \in \mathbb{N}_A} M(p_i) p_i$ is used to denote marking M in $M_L^\star$ or $M_{FBM}^\star$. For an FBM $M \in M_{FBM}^\star$, the operation places that are marked at M are considered only. Thus, marking M' in $\mathcal{L}_M$ is denoted by $\sum_{i \in \mathbb{N}_A, M(p_i) \neq 0} M'(p_i) p_i$.

The Petri net model of an AMS shown in Fig. 5.1 is considered to illustrate the application of the presented deadlock prevention policy. In this Petri net model, there are 11 places and eight transitions. The places can be classified into three groups: idle place set $P^0 = \{p_1, p_8\}$, operation place set $P_A = \{p_2 - p_7\}$, and resource place set $P_R = \{p_9, p_{10}, p_{11}\}$. The net has 28 reachable markings, 19 of which are legal and seven are FBMs. By using the method presented in Section 5.2.1, its reachability graph is analyzed. The minimal covered set of FBMs $M_{FBM}^\star$ and the minimal covering set of legal markings $M_L^\star$ have three and two elements, respectively. Specifically, we have $M_{FBM}^\star = \{p_2 + p_5, p_3 + p_5, p_2 + p_6\}$ and $M_L^\star = \{p_5 + p_6 + p_7, 2p_2 + p_3 + p_4\}$.

At the first iteration, FBM$_1 = p_2 + p_5$ is selected. We have $\mathcal{L}_{FBM_1} = \{p_5, 2p_2\}$. In order to forbid FBM$_1$, a PI I_1 is designed to satisfy the constraint: $l_2 \cdot \mu_2 + l_5 \cdot \mu_5 \leq l_2 \cdot 1 + l_5 \cdot 1 - 1$. For the optimal control purposes, I_1 must not forbid any markings in $\mathcal{L}_{FBM_1}$. Thus, I_1 has to satisfy Eq. (5.13): $l_2 \cdot (0 - 1) + l_5 \cdot (1 - 1) \leq -1$ and $l_2 \cdot (2 - 1) + l_5 \cdot (0 - 1) \leq -1$ for the two legal markings p_5 and $2p_2$ in $\mathcal{L}_{FBM_1}$,

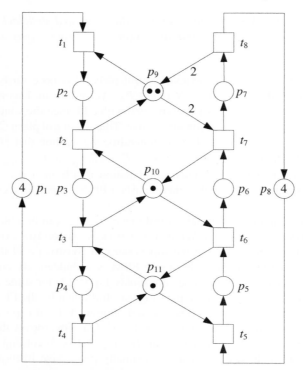

Figure 5.1 Petri net model of a manufacturing system.

respectively. By simplifying the two constraints, we have:

$$-l_2 \leq -1 \quad \text{and}$$

$$l_2 - l_5 \leq -1.$$

The above integer linear system has a solution with $l_2 = 1$ and $l_5 = 2$. Then, a control place p_{c_1} can be designed for I_1: $\mu_2 + 2\mu_5 + \mu_{p_{c_1}} = 2$ by the method presented in Section 5.2.1. As a result, we have $M_0(p_{c_1}) = 2$, $^\bullet p_{c_1} = \{t_2, 2t_6\}$, and $p_{c_1}^\bullet = \{t_1, 2t_5\}$. Only FBM$_1$ is forbidden by I_1, i.e., $F_{I_1} = \{\text{FBM}_1\}$. Hence, we have $\mathcal{M}_{\text{FBM}}^\star := \mathcal{M}_{\text{FBM}}^\star - \{\text{FBM}_1\}$, i.e., $\mathcal{M}_{\text{FBM}}^\star = \{p_3 + p_5, p_2 + p_6\}$.

At the second iteration, FBM$_2 = p_3 + p_5$ is selected. We have $\mathcal{L}_{\text{FBM}_2} = \{p_5, p_3\}$. In order to forbid FBM$_2$, I_2 is designed to satisfy the constraint: $l_3 \cdot \mu_3 + l_5 \cdot \mu_5 \leq l_3 \cdot 1 + l_5 \cdot 1 - 1$. For the optimal control purposes, I_2 must not forbid any markings in $\mathcal{L}_{\text{FBM}_2}$. Thus, I_2 has to satisfy Eq. (5.13): $l_3 \cdot (0 - 1) + l_5 \cdot (1 - 1) \leq -1$ and $l_3 \cdot (1 - 1) + l_5 \cdot (0 - 1) \leq -1$ for the two legal markings p_5 and p_3 in $\mathcal{L}_{\text{FBM}_2}$, respectively. By simplifying the two constraints, we have:

$$-l_3 \leq -1 \quad \text{and}$$

$$-l_5 \leq -1.$$

The above integer linear system has a solution with $l_3 = 1$ and $l_5 = 1$. Then, a control place p_{c_2} can be designed for I_2: $\mu_3 + \mu_5 + \mu_{p_{c_2}} = 1$ by the method presented in Section 5.2.1. As a result, we have $M_0(p_{c_2}) = 1$, $^\bullet p_{c_2} = \{t_3, t_6\}$, and $p_{c_2}^\bullet = \{t_2, t_5\}$. Only FBM$_2$ is forbidden by I_2, i.e., $F_{I_2} = \{$FBM$_2\}$. Hence, we have $\mathcal{M}_{FBM}^\star := \mathcal{M}_{FBM}^\star - \{FBM_2\}$, i.e., $\mathcal{M}_{FBM}^\star = \{p_2 + p_6\}$.

Now consider the last FBM, namely FBM$_3 = p_2 + p_6$. We have $\mathcal{L}_{FBM_3} = \{p_6, 2p_2\}$. In order to forbid FBM$_3$, I_3 is designed to satisfy the constraint: $l_2 \cdot \mu_2 + l_6 \cdot \mu_6 \leq l_2 \cdot 1 + l_6 \cdot 1 - 1$. For the optimal control purposes, I_3 must not forbid any markings in $\mathcal{L}_{FBM_3}$. Thus, I_3 also has to satisfy Eq. (5.13): $l_2 \cdot (0-1) + l_6 \cdot (1-1) \leq -1$ and $l_2 \cdot (2-1) + l_6 \cdot (0-1) \leq -1$ for the two legal markings p_6 and $2p_2$ in $\mathcal{L}_{FBM_3}$, respectively. By simplifying the two constraints, we have:

$$-l_2 \leq -1 \quad \text{and}$$

$$l_2 - l_6 \leq -1.$$

The above integer linear system has a solution with $l_2 = 1$ and $l_6 = 2$. Then, a control place p_{c_3} can be designed for I_3: $\mu_2 + 2\mu_6 + \mu_{p_{c_3}} = 2$. As a result, we have $M_0(p_{c_3}) = 2$, $^\bullet p_{c_3} = \{t_2, 2t_7\}$, and $p_{c_3}^\bullet = \{t_1, 2t_6\}$. Only FBM$_3$ is forbidden by I_3, i.e., $F_{I_3} = \{$FBM$_3\}$. Hence, we have $\mathcal{M}_{FBM}^\star := \mathcal{M}_{FBM}^\star - \{FBM_3\}$, i.e., $\mathcal{M}_{FBM}^\star = \emptyset$.

Now, Algorithm 5.3 terminates and three control places are obtained. Table 5.1 shows the results, where the first column is the iteration number i, the second represents the selected first-met bad marking FBM$_i$, the third, N_{LP}, is the number of inequalities to determine PI I_i for FBM$_i$, and the fourth shows I_i by solving the sets of inequalities for FBM$_i$. The fifth to seventh columns indicate pre-transitions ($^\bullet p_{c_i}$), post-transitions ($p_{c_i}^\bullet$), and initial marking ($M_0(p_{c_i})$) of control place p_{c_i}, respectively. Adding the three control places to the original net model, as shown in Fig. 5.2, the resulting net is live with 19 legal markings.

Table 5.1 Control places computed for the net shown in Fig. 5.1

i	FBM$_i$	N_{LP}	I_i	$^\bullet p_{c_i}$	$p_{c_i}^\bullet$	$M_0(p_{c_i})$
1	$p_2 + p_5$	2	$\mu_2 + 2\mu_5 \leq 2$	$t_2, 2t_6$	$t_1, 2t_5$	2
2	$p_3 + p_5$	2	$\mu_3 + \mu_5 \leq 1$	t_3, t_6	t_2, t_5	1
3	$p_2 + p_6$	2	$\mu_2 + 2\mu_6 \leq 2$	$t_2, 2t_7$	$t_1, 2t_6$	2

According to this example, it can be seen that the minimal covered set of FBMs and the minimal covering set of legal markings have three and two markings, respectively. That is to say, we need to solve at most three sets of inequalities and each set has two inequalities only. Two of 19 legal markings and three of seven FBMs are considered only by the presented policy. The two instances are far smaller than those of FBMs and legal markings. Thus, the presented policy can obviously alleviate the computational burden.

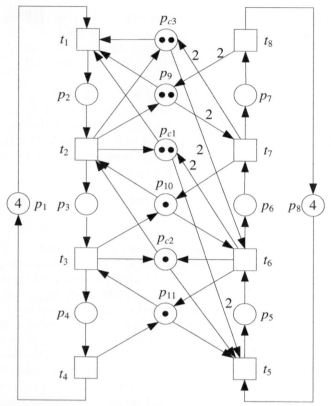

Figure 5.2 An optimally controlled system of the net shown in Fig. 5.1.

5.6 EXPERIMENTAL RESULTS

This section provides experimental results of the policy presented in Section 5.5. For the application of BDDs, C++ program has been used with the BDD tool Buddy-2.2 package (Lind-Nielsen, 2002) on a Linux operating system with Intel CPU Core 2.8 GHz and 4 GB memory.

First, we consider an example from (Park and Reveliotis, 2001), as shown in Fig. 5.3. The net model has the following place partitions: $P^0 = \{p_{10}, p_{20}\}$, $P_R = \{p_{31}, p_{32}, p_{33}\}$, and $P_A = \{p_{11}, \ldots, p_{15}, p_{21}, \ldots, p_{24}\}$. It has 363 reachable markings, 38 and 323 of which are FBMs and legal, respectively. By using the vector covering approach, $\mathcal{M}_{FBM}^\star$ and $\mathcal{M}_L^\star$ have 3 and 48 markings, respectively. Table 5.2 shows the application of the presented policy to this example. There are three control places to be added and the resulting net has all 323 legal markings that represent the maximally permissive behavior. That is to say, the presented policy can lead to a liveness-enforcing supervisor with maximally permissive behavior. If the theory of regions is used, this net model has 40 MTSIs. In this case, we have to solve 40 sets of inequalities and each set has more than 323 inequalities. However, for the method presented in this chapter, the numbers of markings in the minimal covered set of FBMs and the minimal covering set of legal markings are three and 48, respectively. This means that we need to solve at most three sets of inequalities and each set has no more than 48 inequalities. In fact, the biggest set of inequalities has four inequalities only at the third iteration, which is much smaller than 48.

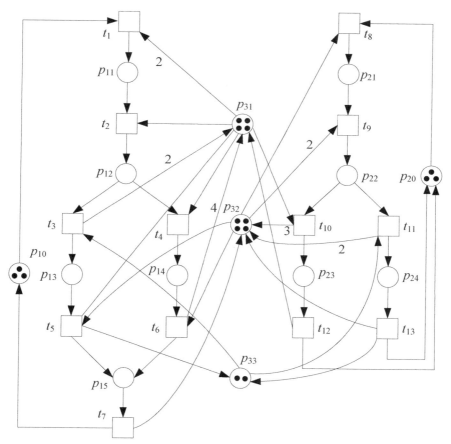

Figure 5.3 An S^3PGR^2 model from (Park and Reveliotis, 2001).

Table 5.2 Control places computed for the net model in Fig. 5.3

i	FBM_i	N_{LP}	I_i	${}^\bullet p_{c_i}$	$p_{c_i}^\bullet$	$M_0(p_{c_i})$
1	$2p_{11}$	1	$\mu_{11} \le 1$	t_2	t_1	1
2	$2p_{21}$	1	$\mu_{21} \le 1$	t_9	t_8	1
3	$p_{11}+2p_{13}+p_{21}+p_{22}$	4	$\mu_{11}+\mu_{13}+\mu_{21}+2\mu_{22} \le 5$	$t_2,t_5,2t_{10},2t_{11}$	t_1,t_3,t_8,t_9	4

Table 5.6 shows the performance comparison of some deadlock control policies in the literature for this example. Note that the policy in (Uzam and Zhou, 2006) leads to 322 reachable markings with three control places, which is suboptimal for

the net model since one legal marking cannot be reached. The presented method leads to an optimal supervisor with all 323 legal markings.

Table 5.3 Performance comparison of typical deadlock control policies for the Petri net in Fig. 5.3

Parameters	(Park and Reveliotis, 2001)	(Uzam and Zhou, 2006)	Alg. 5.3
No. monitors	3	3	3
No. states	85	322	323

Next, we consider the Petri net model of an AMS (Uzam, 2002), as shown in Fig. 5.4. There are 19 places and 14 transitions. It has the following place partitions: $P^0 = \{p_1, p_8\}$, $P_R = \{p_{14}, \dots, p_{19}\}$, and $P_A = \{p_2, \dots, p_7, p_9, \dots, p_{13}\}$. It has 282 reachable markings, 54 and 205 of which are FBMs and legal, respectively. Using the vector covering approach, $\mathcal{M}^\star_{\text{FBM}}$ and $\mathcal{M}^\star_L$ have eight and 26 markings, respectively. Table 5.4 shows the application of the presented policy to the example. The reported method can obtain a maximally permissive supervisor by solving at most eight sets of inequalities and each set has no more than five inequalities. There are eight control places that need to be added and the resulting net has all 205 legal markings. If the theory of regions is used, it is necessary to solve at most 59 sets of inequalities and each set has more than 205 inequalities, as done in (Uzam, 2002).

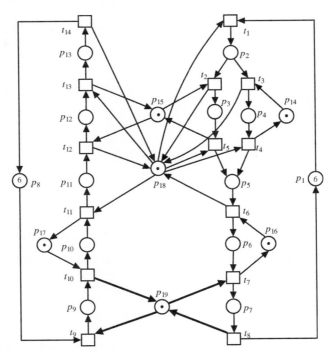

Figure 5.4 Petri net model of an FMS.

Table 5.4 Control places computed for the net model in Fig. 5.4

i	FBM_i	N_{LP}	I_i	$^\bullet p_{c_i}$	$p_{c_i}^\bullet$	$M_0(p_{c_i})$
1	$p_3 + p_{11}$	2	$\mu_3 + \mu_{11} \leq 1$	t_5, t_{12}	t_2, t_{11}	1
2	$p_{11} + p_{12}$	2	$\mu_{11} + \mu_{12} \leq 1$	t_{13}	t_{11}	1
3	$p_2 + p_3 + p_4$	3	$\mu_2 + \mu_3 + \mu_4 \leq 2$	t_4, t_5	t_1	2
4	$p_2 + p_4 + p_{12}$	3	$\mu_2 + \mu_4 + \mu_{12} \leq 2$	t_2, t_4, t_{13}	t_1, t_{12}	2
5	$p_5 + p_6 + p_9 + p_{10}$	4	$\mu_5 + \mu_6 + \mu_9 + \mu_{10} \leq 3$	t_7, t_{11}	t_4, t_5, t_9	3
6	$p_3 + p_6 + p_9 + p_{10}$	4	$\mu_3 + \mu_6 + \mu_9 + \mu_{10} \leq 3$	t_5, t_7, t_{11}	t_2, t_6, t_9	3
7	$p_3 + p_5 + p_9 + p_{10}$	4	$\mu_3 + \mu_5 + \mu_9 + \mu_{10} \leq 3$	t_6, t_{11}	t_2, t_4, t_9	3
8	$p_2 + p_4 + p_6 + p_9 + p_{10}$	5	$\mu_2 + \mu_4 + \mu_6 + \mu_9 + \mu_{10} \leq 4$	t_2, t_4, t_7, t_{11}	t_1, t_6, t_9	4

To further show the advantage of this approach, we vary the initial markings of resource places p_{15}, p_{18}, and p_{19} and idle places p_1 and p_8. Table 5.5 shows various parameters in the net, where the first column represents the initial tokens in places p_1, p_8, p_{15}, p_{18}, and p_{19}; $N_R(nodes)$, N_L, N_{FBM}, $N_L^\star$, and $N_{FBM}^\star$ indicate the numbers of the reachable markings (BDD nodes for representation of reachable markings), legal markings, FBMs, the minimal covering set of legal markings, and the minimal covered set of FBMs, respectively. The seventh and eighth columns are $r_L = N_L^\star / N_L$ and $r_{FBM} = N_{FBM}^\star / N_{FBM}$, respectively. The ninth column shows the total CPU time τ for computing $R(N, M_0)$, $\mathcal{M}_L$, $\mathcal{M}_{FBM}$, $\mathcal{M}_L^\star$, and $\mathcal{M}_{FBM}^\star$ by using BDDs. The number of MTSIs is provided in the final column.

From Table 5.5, it can be seen that when the number of the reachable markings is not big (the first six rows), the number of BDD nodes is bigger than that of reachable markings. In this case, the BDDs and other approaches (for example, using the Petri net analysis software INA (Starke, 2003)) can compute the reachable markings in a short time. When the number of reachable markings increases (the last two rows), we can see that the number of BDD nodes is smaller than that of reachable markings. In this case, the BDD approach becomes more efficient.

In the theory of regions, N_{sep} and N_L in Table 5.5 represent the numbers of the sets of inequalities and inequalities in each set, respectively. However, in the presented policy, the numbers of the two instances are $N_{FBM}^\star$ and no greater than $N_L^\star$, respectively. They are much smaller than their corresponding ones in the theory of regions. The rates r_L and r_{FBM} indicate that only two much smaller subsets of legal markings and FBMs require to be considered. Note that the larger the initial marking is, the smaller the two rates are. This means that the presented method becomes more efficient when handling Petri nets with a large reachability space.

Table 5.5 Parameters in the model depicted in Fig. 5.4 with varying markings

$p_1, p_8, p_{15}, p_{18}, p_{19}$	$N_R(nodes)$	N_L	N_{FBM}	$N_L^\star$	$N_{FBM}^\star$	$r_L/\%$	$r_{FBM}/\%$	τ/S	N_{sep}
6, 6, 1, 1, 1	282(2372)	205	54	26	8	12.7	14.8	< 1	59
7, 6, 2, 1, 1	600(4016)	484	83	43	10	8.9	12.0	< 1	95
7, 6, 1, 2, 1	972(4427)	870	82	70	11	9.4	13.4	1	103
7, 6, 1, 1, 2	570(3576)	421	102	36	8	8.6	7.8	< 1	107
9, 8, 2, 2, 2	4011(14683)	3711	234	147	16	4.0	6.8	3	288
12, 11, 3, 3, 3	27152(42849)	26316	684	496	37	1.9	5.4	25	886
15, 14, 4, 4, 4	124110(100502)	122235	1585	1270	71	1.0	4.5	159	2115
18, 17, 5, 5, 5	440850(251059)	437190	3168	2760	121	0.6	3.8	1234	4311

Consider another example that is taken from (Ezpeleta *et al.*, 1995), as shown in Fig. 5.5. The net model has the following place partitions: $P^0 = \{p_1, p_5, p_{14}\}$, $P_R = \{p_{20}, \ldots, p_{26}\}$, and $P_A = \{p_2, \ldots, p_4, p_6, \ldots, p_{13}, p_{15}, \ldots, p_{19}\}$. It has 26,750

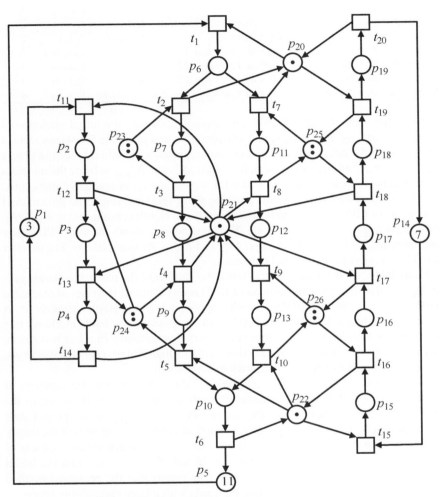

Figure 5.5 An S^3PR model in (Ezpeleta *et al.*, 1995).

reachable markings (represented by 59,255 BDD nodes), 4,211 and 21,581 of which are FBMs and legal, respectively. By using the vector covering approach, $\mathcal{M}^\star_{\text{FBM}}$ and $\mathcal{M}^\star_L$ have 34 and 393 markings, respectively. All the computation can be finished within 30 CPU seconds. Table 5.6 shows the application of the presented policy to this example. There are 17 control places to be added and the resulting net has all 21,581 legal markings that represent the maximally permissive behavior. It is obvious that the presented policy can lead to a liveness-enforcing supervisor with maximally permissive behavior. If the theory of regions is used, this net model has 5,299 MTSIs. That is to say, we have to solve 5,299 sets of inequalities and each set has more than 21,581 inequalities. However, for the presented method, the numbers of markings in the minimal covered set of FBMs and the minimal covering set of legal markings are 34 and 393, respectively. This means that we need to solve at most 34 sets of inequalities and each set has no more than 393 inequalities. In fact, the biggest set of inequalities has 46 inequalities at the 15th iteration step, which is much smaller than 393. Furthermore, we use BDDs to do the required analysis of the reachability space of the net model, which is rather efficient. Table 5.7 shows the performance comparison of a few deadlock control policies in the literature for this example. Note that the policy in (Uzam and Zhou, 2006) leads to 21,562 reachable markings with 19 control places, which is suboptimal for the net model. In (Piroddi *et al.*, 2009), a method that avoids an explicit enumeration of reachable markings and siphons also leads to all 21,581 legal markings for the net with 13 control places in less than 15 seconds, which is rather efficient. The limitation of the method in (Piroddi *et al.*, 2009) is that it can be applied to PT-ordinary nets only.

In order to show the advantage of the presented method, we consider a large example that is taken from (Li and Zhou, 2008) as shown in Fig. 5.6. This model is an S^3PR and has five processes. It has the following place partitions: $P^0 = \{p_1, p_{10}, p_{16}, p_{22}, p_{31}\}$, $P_R = \{p_{37}, \ldots, p_{48}\}$, and $P_A = \{p_2, \ldots, p_9, p_{11}, \ldots, p_{15}, p_{17}, \ldots, p_{21}, p_{23}, \ldots, p_{30}, p_{32}, \ldots, p_{36}\}$. For this net, an explicit enumeration of reachable markings can be obtained in about two days with 142,865,280 reachable markings that can be represented by 5,398,328 BDD nodes. By using Algorithms 3.3 and 3.4, it has 40,319,764 FBMs (represented by 5,405,724 BDD nodes) and 84,489,428 legal markings (represented by 6,183,694 BDD nodes) and the two algorithms can be finished in about three days. Using the vector covering approach, $\mathcal{M}_{FBM}^{\star}$ and $\mathcal{M}_{L}^{\star}$ have 253 and 64,629 markings, respectively. These two processes can be finished in about 14 hours. Table 5.8 shows the application of the presented policy. There are total 53 control places to be computed and this process can be finished in three minutes. It can be seen that the biggest set of inequalities has 322 inequalities at the sixth and 16th iteration steps, which is much smaller than 64,629. Therefore, once the two minimal sets of a model are given, a maximally permissive supervisor can be computed very efficiently. For this example, we can see that a large amount of markings can be represented by a much smaller number of BDD nodes, which indicates that BDDs are efficient to compute the reachable markings of large-scale Petri nets.

Table 5.6 Control places computed for the net model in Fig. 5.5

i	FBM$_i$	N_{LP}	I_i	$\bullet p_{c_i}$	$p_{c_i}^\bullet$	$M_0(p_{c_i})$
1	$p_2 + 2p_3$	2	$\mu_2 + \mu_3 \leq 2$	t_{13}	t_{11}	2
2	$2p_3 + p_8$	2	$\mu_3 + \mu_8 \leq 2$	t_4, t_{13}	t_3, t_{12}	2
3	$2p_{11} + p_{17}$	2	$\mu_{11} + \mu_{17} \leq 2$	t_8, t_{18}	t_7, t_{17}	2
4	$p_{12} + 2p_{16}$	2	$\mu_{12} + \mu_{16} \leq 2$	t_9, t_{17}	t_8, t_{16}	2
5	$2p_{13} + p_{15}$	2	$\mu_{13} + \mu_{15} \leq 2$	t_{10}, t_{16}	t_9, t_{15}	2
7	$p_{12} + p_{13} + p_{15} + p_{16}$	6	$\mu_{12} + \mu_{13} + \mu_{15} + \mu_{16} \leq 3$	t_{10}, t_{17}	t_8, t_{15}	3
6	$2p_{11} + 2p_{16}$	2	$\mu_{11} + \mu_{16} \leq 3$	t_8, t_{17}	t_7, t_{16}	3
8	$2p_{11} + p_{12} + p_{15} + p_{16}$	5	$\mu_{11} + \mu_{12} + \mu_{15} + \mu_{16} \leq 4$	t_8, t_{17}	t_7, t_{15}	4
9	$2p_{11} + p_{13} + p_{15} + p_{16}$	6	$\mu_{11} + \mu_{13} + \mu_{15} + \mu_{16} \leq 4$	t_8, t_{10}, t_{17}	t_7, t_9, t_{15}	4
10	$p_2 + p_3 + p_9 + p_{13} + p_{15} + p_{16}$	21	$\mu_2 + \mu_3 + \mu_9 + \mu_{13} + \mu_{15} + \mu_{16} \leq 5$	$t_5, t_{10}, t_{13}, t_{17}$	t_4, t_9, t_{11}, t_{15}	5
11	$p_3 + p_8 + p_9 + p_{13} + p_{15} + p_{16}$	21	$\mu_3 + \mu_8 + \mu_9 + \mu_{13} + \mu_{15} + \mu_{16} \leq 5$	$t_5, t_{10}, t_{13}, t_{17}$	t_3, t_9, t_{12}, t_{15}	5
12	$p_6 + 2p_7 + p_{11} + p_{17} + p_{18}$	9	$\mu_6 + \mu_7 + \mu_{11} + \mu_{17} + \mu_{18} \leq 5$	t_3, t_8, t_{19}	t_1, t_{17}	5
13	$p_2 + p_3 + p_9 + 2p_{11} + p_{15} + p_{16}$	18	$\mu_2 + \mu_3 + \mu_9 + \mu_{11} + \mu_{15} + \mu_{16} \leq 6$	t_5, t_8, t_{13}, t_{17}	t_4, t_7, t_{11}, t_{15}	6
14	$p_3 + p_8 + p_9 + 2p_{11} + p_{15} + p_{16}$	18	$\mu_3 + \mu_8 + \mu_9 + \mu_{11} + \mu_{15} + \mu_{16} \leq 6$	t_5, t_8, t_{13}, t_{17}	t_3, t_7, t_{12}, t_{15}	6
15	$p_6 + 2p_7 + p_8 + p_9 + p_{11} + p_{13} + p_{15} + p_{16} + p_{17}$	46	$\mu_6 + \mu_7 + \mu_8 + \mu_9 + \mu_{11} + 2\mu_{13} + 2\mu_{15} + 2\mu_{16} + \mu_{18} \leq 12$	$t_5, t_8, 2t_{10}, 2t_{17}, t_{19}$	$t_1, 2t_9, 2t_{15}, t_{18}$	12
16	$p_6 + 2p_7 + 2p_9 + p_{11} + p_{13} + p_{15} + p_{16} + p_{17}$	18	$\mu_6 + \mu_7 + \mu_9 + \mu_{11} + \mu_{13} + \mu_{15} + \mu_{16} + \mu_{17} \leq 9$	$t_3, t_5, t_8, t_{10}, t_{18}$	t_1, t_4, t_9, t_{15}	9
17	$p_6 + 2p_7 + 2p_9 + p_{11} + p_{12} + p_{15} + p_{16} + p_{18}$	22	$\mu_6 + \mu_7 + \mu_9 + \mu_{11} + \mu_{12} + \mu_{15} + \mu_{16} + \mu_{18} \leq 9$	$t_3, t_5, t_9, t_{17}, t_{19}$	t_1, t_4, t_{15}, t_{18}	9

Table 5.7 Performance comparison of typical deadlock control policies

Parameters	(Ezpeleta et al., 1995)	(Li and Zhou, 2004)	(Huang et al., 2001)	(Li and Zhou, 2008)	(Uzam and Zhou, 2006)	(Piroddi et al., 2008)	Alg. 5.3
No. monitors	18	6	16	7	19	13	17
No. states	6287	6287	12656	13482	21562	21581	21581

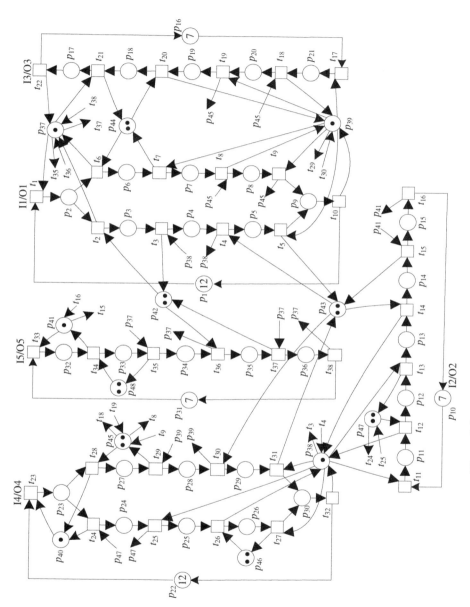

Figure 5.6 A large-sized Petri net model from (Li and Zhou, 2008).

Table 5.8 Control places computed for the net model in Fig. 5.6

i	FBM_i	N_{LP}	I_i	$\bullet p_{c_i}$	$p_{c_i}\bullet$	$M_0(p_{c_i})$
1	$p_2 + p_3 + p_4 + p_5 + p_6 + p_{14} + p_{18} + p_{19} + p_{32} + 2p_{33} + p_{35}$	106	$\mu_2 + \mu_3 + \mu_4 + \mu_5 + \mu_6 + \mu_{14} + \mu_{18} + \mu_{19} + \mu_{32} + \mu_{33} + \mu_{35} \le 11$	$t_5, t_7, t_{15}, t_{21}, t_{35}, t_{37}$	$t_1, t_{14}, t_{19}, t_{33}, t_{36}$	11
2	$p_2 + p_3 + p_4 + p_5 + p_6 + p_{14} + p_{18} + p_{28} + p_{32} + 2p_{33} + p_{35}$	125	$\mu_2 + \mu_3 + \mu_4 + \mu_5 + \mu_6 + \mu_{14} + \mu_{18} + \mu_{28} + \mu_{32} + \mu_{33} + \mu_{35} \le 11$	$t_5, t_7, t_{15}, t_{21}, t_{30}, t_{35}, t_{37}$	$t_1, t_{14}, t_{20}, t_{29}, t_{33}, t_{36}$	11
3	$p_2 + p_3 + p_4 + p_5 + p_6 + p_{18} + p_{19} + p_{29} + p_{35}$	76	$\mu_2 + \mu_3 + \mu_4 + \mu_5 + \mu_6 + \mu_{18} + \mu_{19} + \mu_{29} + \mu_{35} \le 8$	$t_5, t_7, t_{21}, t_{31}, t_{37}$	$t_1, t_{19}, t_{30}, t_{36}$	8
4	$p_2 + p_3 + p_4 + p_6 + p_{14} + p_{18} + p_{19} + p_{29} + p_{32} + 2p_{33} + p_{35}$	88	$\mu_2 + \mu_3 + \mu_4 + \mu_6 + \mu_{14} + \mu_{18} + \mu_{19} + \mu_{29} + \mu_{32} + \mu_{33} + \mu_{35} \le 11$	$t_4, t_7, t_{15}, t_{21}, t_{31}, t_{35}, t_{37}$	$t_1, t_{14}, t_{19}, t_{30}, t_{33}, t_{36}$	11
5	$p_2 + p_3 + p_4 + p_6 + p_{14} + p_{18} + p_{20} + p_{21} + p_{29} + p_{32} + 2p_{33} + p_{35}$	176	$\mu_2 + \mu_3 + \mu_4 + \mu_6 + \mu_{14} + \mu_{18} + \mu_{20} + \mu_{21} + \mu_{29} + \mu_{32} + \mu_{33} + \mu_{35} \le 12$	$t_4, t_7, t_{15}, t_{19}, t_{21}, t_{31}, t_{35}, t_{37}$	$t_1, t_{14}, t_{17}, t_{20}, t_{30}, t_{33}, t_{36}$	12
6	$p_2 + p_3 + p_4 + p_6 + p_{14} + p_{18} + p_{20} + p_{27} + p_{29} + p_{32} + 2p_{33} + p_{35}$	322	$\mu_2 + \mu_3 + \mu_4 + \mu_6 + \mu_{14} + \mu_{18} + \mu_{20} + \mu_{27} + \mu_{29} + \mu_{32} + \mu_{33} + \mu_{35} \le 12$	$t_4, t_7, t_{15}, t_{19}, t_{21}, t_{29}, t_{31}, t_{35}, t_{37}$	$t_1, t_{14}, t_{18}, t_{28}, t_{30}, t_{33}, t_{36}$	12
7	$p_2 + p_3 + p_4 + p_6 + p_{14} + p_{18} + p_{21} + p_{27} + p_{29} + p_{32} + 2p_{33} + p_{35}$	250	$\mu_2 + \mu_3 + \mu_4 + \mu_6 + \mu_{14} + \mu_{18} + \mu_{21} + \mu_{27} + \mu_{29} + \mu_{32} + \mu_{33} + \mu_{35} \le 12$	$t_4, t_7, t_{15}, t_{18}, t_{21}, t_{29}, t_{31}, t_{35}, t_{37}$	$t_1, t_{14}, t_{17}, t_{20}, t_{28}, t_{30}, t_{33}, t_{36}$	12
8	$p_2 + p_3 + p_4 + p_6 + p_{14} + p_{18} + p_{28} + p_{29} + p_{32} + 2p_{33} + p_{35}$	125	$\mu_2 + \mu_3 + \mu_4 + \mu_6 + \mu_{14} + \mu_{18} + \mu_{28} + \mu_{29} + \mu_{32} + \mu_{33} + \mu_{35} \le 11$	$t_4, t_7, t_{15}, t_{21}, t_{31}, t_{35}, t_{37}$	$t_1, t_{14}, t_{20}, t_{29}, t_{33}, t_{36}$	11
9	$p_2 + p_3 + p_4 + p_6 + p_{14} + p_{19} + p_{20} + p_{27} + p_{29} + p_{32} + 2p_{33} + p_{35}$	201	$\mu_2 + \mu_3 + \mu_4 + \mu_6 + \mu_{14} + \mu_{19} + \mu_{20} + \mu_{27} + \mu_{29} + \mu_{32} + \mu_{33} + \mu_{35} \le 12$	$t_4, t_7, t_{15}, t_{20}, t_{29}, t_{31}, t_{35}, t_{37}$	$t_1, t_{14}, t_{18}, t_{28}, t_{30}, t_{33}, t_{36}$	12
10	$p_2 + p_3 + p_4 + p_{14} + 2p_{18} + p_{29} + p_{32} + 2p_{33} + p_{35}$	36	$\mu_2 + \mu_3 + \mu_4 + \mu_{14} + \mu_{18} + \mu_{29} + \mu_{32} + \mu_{33} + \mu_{35} \le 10$	$t_4, t_6, t_{15}, t_{21}, t_{31}, t_{35}, t_{37}$	$t_1, t_{14}, t_{20}, t_{30}, t_{33}, t_{36}$	10
11	$p_2 + p_3 + p_5 + p_6 + p_{13} + p_{14} + p_{18} + p_{19} + p_{32} + 2p_{33} + p_{35}$	106	$\mu_2 + \mu_3 + \mu_5 + \mu_6 + \mu_{13} + \mu_{14} + \mu_{18} + \mu_{19} + \mu_{32} + \mu_{33} + \mu_{35} \le 11$	$t_3, t_5, t_7, t_{15}, t_{21}, t_{35}, t_{37}$	$t_1, t_4, t_{13}, t_{19}, t_{33}, t_{36}$	11
12	$p_2 + p_3 + p_5 + p_6 + p_{13} + p_{14} + p_{18} + p_{28} + p_{32} + 2p_{33} + p_{35}$	125	$\mu_2 + \mu_3 + \mu_5 + \mu_6 + \mu_{13} + \mu_{14} + \mu_{18} + \mu_{28} + \mu_{32} + \mu_{33} + \mu_{35} \le 11$	$t_3, t_5, t_7, t_{15}, t_{21}, t_{30}, t_{35}, t_{37}$	$t_1, t_4, t_{13}, t_{20}, t_{29}, t_{33}, t_{36}$	11
13	$p_2 + p_3 + p_5 + p_6 + p_{13} + p_{18} + p_{19} + p_{29} + p_{35}$	76	$\mu_2 + \mu_3 + \mu_5 + \mu_6 + \mu_{13} + \mu_{18} + \mu_{19} + \mu_{29} + \mu_{35} \le 8$	$t_3, t_5, t_7, t_{14}, t_{21}, t_{31}, t_{37}$	$t_1, t_4, t_{13}, t_{19}, t_{30}, t_{36}$	8
14	$p_2 + p_3 + p_6 + p_{13} + p_{14} + p_{18} + p_{19} + p_{29} + p_{32} + 2p_{33} + p_{35}$	88	$\mu_2 + \mu_3 + \mu_6 + \mu_{13} + \mu_{14} + \mu_{18} + \mu_{19} + \mu_{29} + \mu_{32} + \mu_{33} + \mu_{35} \le 11$	$t_3, t_7, t_{15}, t_{21}, t_{31}, t_{35}, t_{37}$	$t_1, t_{13}, t_{19}, t_{30}, t_{33}, t_{36}$	11
15	$p_2 + p_3 + p_6 + p_{13} + p_{14} + p_{18} + p_{20} + p_{21} + p_{29} + p_{32} + 2p_{33} + p_{35}$	176	$\mu_2 + \mu_3 + \mu_6 + \mu_{13} + \mu_{14} + \mu_{18} + \mu_{20} + \mu_{21} + \mu_{29} + \mu_{32} + \mu_{33} + \mu_{35} \le 12$	$t_3, t_7, t_{15}, t_{19}, t_{21}, t_{31}, t_{35}, t_{37}$	$t_1, t_{13}, t_{17}, t_{20}, t_{30}, t_{33}, t_{36}$	12

Table 5.8 Contd.

Table 5.8 Contd.

i	FBM_i	N_{LP}	I_i	$^\bullet p_{c_i}$	$p_{c_i}^\bullet$	$M_0(p_{c_i})$
16	$p_2 + p_3 + p_6 + p_{13} + p_{14} + p_{18} + p_{20} + p_{27} + p_{29} + p_{32} + p_{33} + p_{35}$	322	$\mu_2 + \mu_3 + \mu_6 + \mu_{13} + \mu_{14} + \mu_{18} + \mu_{20} + \mu_{27} + \mu_{29} + \mu_{32} + \mu_{33} + \mu_{35} \leq 12$	$t_3, t_7, t_{15}, t_{19}, t_{21}, t_{29}, t_{31}, t_{35}, t_{37}$	$t_1, t_{13}, t_{18}, t_{20}, t_{28}, t_{30}, t_{33}, t_{36}$	12
17	$p_2 + p_3 + p_6 + p_{13} + p_{14} + p_{18} + p_{21} + p_{27} + p_{29} + p_{32} + p_{33} + p_{35}$	250	$\mu_2 + \mu_3 + \mu_6 + \mu_{13} + \mu_{14} + \mu_{18} + \mu_{21} + \mu_{27} + \mu_{29} + \mu_{32} + \mu_{33} + \mu_{35} \leq 12$	$t_3, t_7, t_{15}, t_{18}, t_{21}, t_{29}, t_{31}, t_{35}, t_{37}$	$t_1, t_{13}, t_{17}, t_{20}, t_{28}, t_{30}, t_{33}, t_{36}$	12
18	$p_2 + p_3 + p_6 + p_{13} + p_{14} + p_{18} + p_{28} + p_{29} + p_{32} + p_{33} + p_{35}$	125	$\mu_2 + \mu_3 + \mu_6 + \mu_{13} + \mu_{14} + \mu_{18} + \mu_{28} + \mu_{29} + \mu_{32} + \mu_{33} + \mu_{35} \leq 11$	$t_3, t_7, t_{15}, t_{21}, t_{31}, t_{35}, t_{37}$	$t_1, t_{13}, t_{20}, t_{29}, t_{33}, t_{36}$	11
19	$p_2 + p_3 + p_6 + p_{13} + p_{14} + p_{19} + p_{20} + p_{27} + p_{29} + p_{32} + p_{33} + p_{35}$	201	$\mu_2 + \mu_3 + \mu_6 + \mu_{13} + \mu_{14} + \mu_{19} + \mu_{20} + \mu_{27} + \mu_{29} + \mu_{32} + \mu_{33} + \mu_{35} \leq 12$	$t_3, t_7, t_{15}, t_{20}, t_{29}, t_{31}, t_{35}, t_{37}$	$t_1, t_{13}, t_{18}, t_{28}, t_{30}, t_{33}, t_{36}$	12
20	$p_2 + p_3 + p_{13} + p_{14} + 2p_{18} + p_{29} + p_{32} + 2p_{33} + p_{35}$	36	$\mu_2 + \mu_3 + \mu_{13} + \mu_{14} + \mu_{18} + \mu_{29} + \mu_{32} + \mu_{33} + \mu_{35} \leq 10$	$t_3, t_6, t_{15}, t_{21}, t_{31}, t_{35}, t_{37}$	$t_1, t_{13}, t_{20}, t_{30}, t_{33}, t_{36}$	10
21	$p_2 + p_4 + p_6 + p_{14} + p_{18} + p_{20} + p_{27} + p_{28} + p_{32} + 2p_{33} + 2p_{35}$	50	$2\mu_2 + \mu_4 + 2\mu_6 + \mu_{14} + 2\mu_{18} + \mu_{20} + \mu_{27} + \mu_{28} + 2\mu_{32} + 2\mu_{33} + 4\mu_{35} \leq 20$	$2t_2, t_4, 2t_7, t_{15}, t_{19}, 2t_{21}, t_{30}, 2t_{35}, 2t_{37}$	$2t_1, t_3, t_{14}, t_{18}, 2t_{20}, t_{28}, 2t_{33}, 2t_{36}$	20
22	$p_2 + p_5 + p_6 + p_{14} + p_{18} + p_{28} + p_{32} + 2p_{33} + 2p_{35}$	36	$\mu_2 + \mu_5 + \mu_6 + \mu_{14} + \mu_{18} + \mu_{28} + \mu_{32} + \mu_{33} + \mu_{35} \leq 10$	$t_2, t_5, t_7, t_{15}, t_{21}, t_{30}, t_{35}, t_{37}$	$t_1, t_4, t_{14}, t_{20}, t_{29}, t_{33}, t_{36}$	10
23	$p_2 + p_6 + p_{13} + p_{14} + p_{18} + p_{20} + p_{27} + p_{28} + p_{32} + 2p_{33} + 2p_{35}$	50	$2\mu_2 + 2\mu_6 + \mu_{13} + \mu_{14} + 2\mu_{18} + \mu_{20} + \mu_{27} + \mu_{28} + 2\mu_{32} + 2\mu_{33} + 2\mu_{35} \leq 20$	$2t_2, 2t_7, t_{15}, t_{19}, 2t_{21}, t_{30}, 2t_{35}, 2t_{37}$	$2t_1, t_{13}, t_{18}, 2t_{20}, t_{28}, 2t_{33}, 2t_{36}$	20
24	$p_2 + p_6 + 2p_{14} + p_{18} + p_{20} + p_{27} + p_{32} + 2p_{33} + 2p_{35}$	38	$\mu_2 + \mu_6 + \mu_{14} + \mu_{18} + \mu_{20} + \mu_{27} + \mu_{32} + \mu_{33} + \mu_{35} \leq 11$	$t_2, t_7, t_{15}, t_{19}, t_{21}, t_{29}, t_{35}, t_{37}$	$t_1, t_{14}, t_{18}, t_{20}, t_{28}, t_{33}, t_{36}$	11
25	$p_2 + p_6 + 2p_{14} + p_{18} + p_{21} + p_{27} + p_{32} + 2p_{33} + 2p_{35}$	32	$\mu_2 + \mu_6 + \mu_{14} + \mu_{18} + \mu_{21} + \mu_{27} + \mu_{32} + \mu_{33} + \mu_{35} \leq 11$	$t_2, t_7, t_{15}, t_{18}, t_{21}, t_{29}, t_{35}, t_{37}$	$t_1, t_{14}, t_{17}, t_{20}, t_{28}, t_{33}, t_{36}$	11
26	$p_2 + p_6 + 2p_{14} + p_{19} + p_{20} + p_{27} + p_{32} + 2p_{33} + 2p_{35}$	29	$\mu_2 + \mu_6 + \mu_{14} + \mu_{19} + \mu_{20} + \mu_{27} + \mu_{32} + \mu_{33} + \mu_{35} \leq 11$	$t_2, t_7, t_{15}, t_{20}, t_{29}, t_{35}, t_{37}$	$t_1, t_{14}, t_{18}, t_{28}, t_{33}, t_{36}$	11
27	$p_2 + p_6 + p_{18} + p_{19} + 2p_{35}$	8	$\mu_2 + \mu_6 + \mu_{18} + \mu_{19} + \mu_{35} \leq 5$	t_2, t_7, t_{21}, t_{37}	t_1, t_{19}, t_{36}	5
28	$p_2 + p_6 + p_{18} + p_{20} + p_{21} + 2p_{35}$	16	$\mu_2 + \mu_6 + \mu_{18} + \mu_{20} + \mu_{21} + \mu_{35} \leq 6$	$t_2, t_7, t_{19}, t_{21}, t_{37}$	$t_1, t_{17}, t_{20}, t_{36}$	6
29	$p_2 + 2p_{18} + 2p_{35}$	3	$\mu_2 + \mu_{18} + \mu_{35} \leq 4$	t_2, t_6, t_{21}, t_{37}	t_1, t_{20}, t_{36}	4
30	$p_3 + p_4 + p_5 + p_6 + p_{14} + p_{18} + p_{19} + p_{32} + 2p_{33} + p_{34} + p_{35}$	99	$\mu_3 + \mu_4 + \mu_5 + \mu_6 + \mu_{14} + \mu_{18} + \mu_{19} + \mu_{32} + \mu_{33} + \mu_{34} + \mu_{35} \leq 11$	$t_5, t_7, t_{15}, t_{21}, t_{37}$	$t_2, t_6, t_{14}, t_{19}, t_{33}$	11
31	$p_3 + p_4 + p_5 + p_6 + p_{18} + p_{19} + p_{29} + p_{34} + p_{35}$	67	$\mu_3 + \mu_4 + \mu_5 + \mu_6 + \mu_{18} + \mu_{19} + \mu_{29} + \mu_{34} + \mu_{35} \leq 8$	$t_5, t_7, t_{21}, t_{31}, t_{37}$	$t_2, t_6, t_{19}, t_{30}, t_{35}$	8
32	$p_3 + p_4 + p_5 + p_{14} + p_{28} + p_{32} + 2p_{33} + p_{34} + p_{35}$	31	$\mu_3 + \mu_4 + \mu_5 + \mu_{14} + \mu_{28} + \mu_{32} + \mu_{33} + \mu_{34} + \mu_{35} \leq 9$	$t_5, t_{15}, t_{30}, t_{37}$	$t_2, t_{14}, t_{29}, t_{33}$	9

Table 5.8 Contd.

Table 5.8 Contd.

i	FBM_i	N_{LP}	I_i	$\bullet p_{c_i}$	$p_{c_i}^\bullet$	$M_0(p_{c_i})$
33	$p_3 + p_4 + p_{14} + p_{29} + p_{32} + p_{33} + p_{34} + p_{35}$	29	$\mu_3 + \mu_4 + \mu_{14} + \mu_{29} + \mu_{32} + \mu_{33} + \mu_{34} + \mu_{35} \le 8$	$t_4, t_{15}, t_{31}, t_{37}$	$t_2, t_{14}, t_{30}, t_{33}$	8
34	$p_3 + p_5 + p_6 + p_{13} + p_{14} + p_{18} + p_{19} + p_{32} + 2p_{33} + p_{34} + p_{35}$	99	$\mu_3 + \mu_5 + \mu_6 + \mu_{13} + \mu_{14} + \mu_{18} + \mu_{19} + \mu_{32} + \mu_{33} + \mu_{34} + \mu_{35} \le 11$	$t_3, t_5, t_7, t_{15}, t_{21}, t_{37}$	$t_2, t_4, t_6, t_{13}, t_{19}, t_{33}$	11
35	$p_3 + p_5 + p_6 + p_{13} + p_{18} + p_{19} + p_{29} + p_{34} + p_{35}$	67	$\mu_3 + \mu_5 + \mu_6 + \mu_{13} + \mu_{18} + \mu_{19} + \mu_{29} + \mu_{34} + \mu_{35} \le 8$	$t_3, t_5, t_7, t_{14}, t_{21}, t_{31}, t_{37}$	$t_2, t_4, t_6, t_{13}, t_{19}, t_{30}, t_{35}$	8
36	$p_3 + p_5 + p_{13} + p_{14} + p_{28} + p_{32} + 2p_{33} + p_{35}$	31	$\mu_3 + \mu_5 + \mu_{13} + \mu_{14} + \mu_{28} + \mu_{32} + \mu_{33} + \mu_{34} + \mu_{35} \le 9$	$t_3, t_5, t_{15}, t_{30}, t_{37}$	$t_2, t_4, t_{13}, t_{29}, t_{33}$	9
37	$p_3 + p_{13} + p_{14} + p_{29} + p_{32} + 2p_{33} + p_{34} + p_{35}$	29	$\mu_3 + \mu_{13} + \mu_{14} + \mu_{29} + \mu_{32} + \mu_{33} + \mu_{34} + \mu_{35} \le 8$	$t_3, t_{15}, t_{31}, t_{37}$	$t_2, t_{13}, t_{30}, t_{33}$	8
38	$p_4 + p_5 + p_{28} + p_{29}$	6	$\mu_4 + \mu_5 + \mu_{28} + \mu_{29} \le 3$	t_5, t_{31}	t_3, t_{29}	3
39	$p_4 + 2p_{29}$	2	$\mu_4 + \mu_{29} \le 2$	t_4, t_{31}	t_3, t_{30}	2
40	$p_5 + p_{13} + p_{28} + p_{29}$	6	$\mu_5 + \mu_{13} + \mu_{28} + \mu_{29} \le 3$	t_5, t_{14}, t_{31}	t_4, t_{13}, t_{29}	3
41	$2p_5 + p_{28}$	2	$\mu_5 + \mu_{28} \le 2$	t_5, t_{30}	t_4, t_{29}	2
42	$2p_6 + p_{19}$	2	$\mu_6 + \mu_{19} \le 2$	t_7, t_{20}	t_6, t_{19}	2
43	$2p_6 + p_{20} + p_{21}$	4	$\mu_6 + \mu_{20} + \mu_{21} \le 3$	t_7, t_{19}	t_6, t_{17}	3
44	$p_7 + p_8 + p_{20}$	5	$\mu_7 + \mu_8 + \mu_{20} \le 2$	t_9, t_{19}	t_7, t_{18}	2
45	$p_7 + p_8 + p_{27}$	5	$\mu_7 + \mu_8 + \mu_{27} \le 2$	t_9, t_{29}	t_7, t_{28}	2
46	$p_7 + p_{20} + p_{27}$	5	$\mu_7 + \mu_{20} + \mu_{27} \le 2$	t_8, t_{19}, t_{29}	t_7, t_{18}, t_{28}	2
47	$p_8 + p_{20} + p_{21}$	5	$\mu_8 + \mu_{20} + \mu_{21} \le 2$	t_9, t_{19}	t_8, t_{17}	2
48	$p_8 + p_{21} + p_{27}$	5	$\mu_8 + \mu_{21} + \mu_{27} \le 2$	t_9, t_{18}, t_{29}	t_8, t_{17}, t_{28}	2
49	$p_{11} + p_{12} + p_{24}$	5	$\mu_{11} + \mu_{12} + \mu_{24} \le 2$	t_{13}, t_{25}	t_{11}, t_{24}	2
50	$p_{13} + 2p_{29}$	2	$\mu_{13} + \mu_{29} \le 2$	t_{14}, t_{31}	t_{13}, t_{30}	2
51	$p_{20} + p_{21} + p_{27}$	5	$\mu_{20} + \mu_{21} + \mu_{27} \le 2$	t_{19}, t_{29}	t_{17}, t_{28}	2
52	$p_{25} + 2p_{26}$	2	$\mu_{25} + \mu_{26} \le 2$	t_{27}	t_{25}	2
53	$p_{34} + 2p_{35}$	2	$\mu_{34} + \mu_{35} \le 2$	t_{37}	t_{35}	2

Note that in (Li and Zhou, 2008), an elementary siphon method obtains a supervisor with 10 control places in about eight minutes (since a complete enumeration of all strict minimal siphons can be achieved in eight minutes). Another very typical result for the example is presented by the selective siphon approach in (Piroddi *et al.*, 2009). In this study, Piroddi *et al.* derive a supervisor with 25 control places in 90 seconds, which is rather efficient. Both of the methods are more efficient than the presented method in the sense of computational time. However, the elementary siphon approach results in a live controlled net with only 257,890 reachable markings that are much less than the legal markings. The selective siphon approach is possible to obtain a maximally permissive supervisor. Unfortunately, maximal permissiveness is not formally proved in (Piroddi *et al.*, 2009).

5.7 CONCLUSIONS

First, we discuss the complexity of the presented method. The problem under consideration is NP-hard since one requires a complete generation of reachable markings and needs to solve ILPPs, both of which are NP-hard problems. In order to overcome the high computational cost of the method, two techniques are provided in this monograph: a vector covering approach and BDDs for the generation of a reachability graph (see Chapter 3).

The vector covering approach can reduce the legal markings and FBMs into two very small sets, namely the minimal covering set of legal markings and the minimal covered set of FBMs. When an FBM is selected, a further vector covering approach is applied and the minimal covering set of legal markings is reduced into a further smaller one. The main contribution of this technique is that it greatly reduces the number of the sets of inequalities and the number of inequalities in each set. It is known that an ILPP is NP-hard. However, an ILPP with only a small number of constraints can be solved efficiently. It can be seen from the experimental results that even for a large example that has more than 80 million of legal markings, the largest number of inequalities for each set is 322 only. Thus, the vector covering approach can greatly reduce the computational burden of solving an ILPP and make the presented method applicable for large-scale Petri nets.

To overcome the inefficiency of explicitly enumerating reachable markings by a traditional computation method, the application of BDDs (Andersen, 1997; Brant, 1992; Miner and Ciardo, 1999; Pastor *et al.*, 1994, 2001) to Petri net analysis is introduced. Chapter 3 develops a number of methods to compute the sets of legal markings and FBMs by using BDDs. Experimental results show that BDDs are effective and efficient to compute a reachability graph. For the net of the large example, an explicit enumeration of reachable markings is impractical. However, BDDs can be used to overcome this problem. In this point, BDDs are powerful to compute reachable markings.

Although the complexity of the problem under consideration is NP-hard in theory, experimental results show that these two techniques make the presented method feasible to tackle large-scale Petri nets. Thus, from the experimental point of view, the method is still efficient.

Second, we show comparison among relative studies. Uzam and Zhou (2006) develop an effective method to deal with deadlocks in AMSs. They provide a number of examples to illustrate that their method can obtain a highly permissive but not, in general, optimal supervisor. Ghaffari *et al.* (2003) develop a method using the theory of regions, which is an optimal but computationally expensive approach since a lot of sets of inequalities need to be solved. Both the theory of regions and presented method can obtain a maximally permissive supervisor if such a supervisor exists. However, in the presented method, a vector covering approach is used to reduce the number of the sets of inequalities and the number of inequalities in each set. Thus, compared with the theory of regions, the presented method has a much low computational overhead in practice, which is shown by the experimental study. Compared with the method proposed by Uzam and Zhou (2006), the one presented in this chapter computes the reachability graph of a Petri net only once. Furthermore, it is maximally permissive but Uzam and Zhou's is not. The application scope of this method is limited to bounded Petri net models of AMSs as in Uzam and Zhou's method but smaller than the theory of regions (Ghaffari *et al.*, 2003; Uzam, 2002).

The sequence of FBMs to be chosen at each iteration strongly influences the final control places. A good sequence can lead to a supervisor without redundant control places. Different choices have different iterations times and different redundant control places. The presented method cannot automatically remove or identify the redundant control places from the final supervisor. Interesting work includes defining better rules to choose an FBM at each iteration to reduce iteration times and possible redundant control places.

5.8 BIBLIOGRAPHICAL REMARKS

Most materials of this chapter can be found in (Chen *et al.*, 2011). Other efficient approaches that are applicable to large-scale Petri net models can be found in (Li and Zhou, 2004, 2008) and Piroddi *et al.*'s work (Piroddi *et al.*, 2008, 2009). The well-known control place synthesis by a PI is proposed in (Yamalidou *et al.*, 1996). The concept of FBMs is developed in (Uzam, 2002; Uzam and Zhou, 2007). The classification of places for the Petri net models of AMSs can be found in (Ezpeleta *et al.*, 1995; Park and Reveliotis, 2001; Tricas and Martinez, 1995; Zhou and DiCesare, 1991).

References

Andersen, H. R. 1997. An introduction to binary decision diagrams. Lecture Notes for 49285 Advanced Algorithms E97, Department of Information Technology, Technical University of Denmark.

Banaszak, Z. and B. H. Krogh. 1990. Deadlock avoidance in flexible manufacturing systems with concurrently competing process flows. IEEE Transactions on Robotics and Automation. 6(6): 724–734.

Brant, R. 1992. Symbolic Boolean manipulation with ordered binary decision diagrams. ACM Computing Surveys. 24(3): 293–318.

Chen, Y. F., Z. W. Li, M. Khalgui, and O. Mosbahi. 2011. Design of a maximally permissive liveness-enforcing Petri net supervisor for flexible manufacturing systems. IEEE Transactions on Automation Science and Engineering. 8(2): 374–393.

Cordone, R., L. Ferrarini, and L. Piroddi. 2005. Enumeration algorithms for minimal siphons in petri nets based on place constraints. IEEE Transactions on System, Man and Cybernetics, Part A. 35(6): 844–854.

Ezpeleta, J., J. M. Colom, and J. Martinez. 1995. A Petri net based deadlock prevention policy for flexible manufacturing systems. IEEE Transactions on Robotics and Automation. 11(2): 173–184.

Ezpeleta, J., F. Tricas, F. Garcia-Valles, and J. M. Colom. 2002. A banker's solution for deadlock avoidance in FMS with flexible routing and multiresource states. IEEE Transactions on Robotics and Automation. 18(4): 621–625.

Ghaffari, A., N. Rezg, and X. L. Xie. 2003. Design of a live and maximally permissive Petri net controller using the theory of regions. IEEE Transactions on Robotics and Automation. 19(1): 137–142.

Hsieh, F. S. and S. C. Chang. 1994. Dispatching-driven deadlock avoidance controller synthesis for flexible manufacturing systems. IEEE Transactions on Robotics and Automation. 10(2): 196–209.

Huang, Y. S., M. D. Jeng, X. L. Xie, and S. L. Chung. 2001. Deadlock prevention based on Petri nets and siphons. International Journal of Production Research. 39(2): 283–305.

Huang, Y. S., M. D. Jeng, X. L.Xie, and D. H. Chung. 2006. Siphon-based deadlock prevention for flexible manufacturing systems. IEEE Transactions on Systems, Man, and Cybernetics, Part A. 36(6): 1248–1256.

Lautenbach, K. and H. Ridder. 1994. Liveness in bounded Petri nets which are covered by T-invariants. Lecture Notes in Computer Science. 815: 358–375.

Li, Z. W. and M. C. Zhou. 2004. Elementary siphons of Petri nets and their application to deadlock prevention in flexible manufacturing systems. IEEE Transactions on Systems, Man, and Cybernetics, Part A. 34(1): 38–51.

Li, Z. W. and M. C. Zhou. 2008. Control of elementary and dependent siphons in Petri nets and their application. IEEE Transactions on Systems, Man, and Cybernetics, Part A. 38(1): 133–148.

Li, Z. W., M. C. Zhou, and M. D. Jeng. 2008. A maximally permissive deadlock prevention policy for FMS based on Petri net siphon control and the theory of regions. IEEE Transactions on Automation Science and Engineering. 5(1): 182–188.

Lind-Nielsen, J. 2002. BuDDy: Binary Decision Diagram Package Release 2.2. IT-University of Copenhagen (ITU).

Miner, A. S. and G. Ciardo. 1999. Efficient reachability set generation and storage using decision diagrams. Lecture Notes in Computer Science. 1639: 6–25.

Murata, T. 1989. Petri nets: Properties, analysis and application. Proceedings of the IEEE. 77(4): 541–580.

Park, J. and S. A. Reveliotis. 2000. Algebraic synthesis of efficient deadlock avoidance policies for sequential resource allocation systems. IEEE Transactions on Robotics and Automation. 16(2): 190–195.

Park, J. and S. A. Reveliotis. 2001. Deadlock avoidance in sequential resource allocation systems with multiple resource acquisitions and flexible routings. IEEE Transactions on Automatic Control. 46(10): 1572–1583.

Pastor, E., O. Roig, J. Cortadella, and R. M. Badia. 1994. Petri net analysis using Boolean manipulation. Lecture Notes in Computer Science. 815: 416–435.

Pastor, E., J. Cortadella, and O. Roig. 2001. Symbolic analysis of bounded Petri nets. IEEE Transactions on Computers. 50(5): 432–448.

Piroddi, L., R. Cordone, and I. Fumagalli. 2008. Selective siphon control for deadlock prevention in Petri nets. IEEE Transactions on Systems, Man, and Cybernetics, Part A. 38(6): 1337–1348.

Piroddi, L., R. Cordone, and I. Fumagalli. 2009. Combined siphon and marking generation for deadlock prevention in Petri nets. IEEE Transactions on Systems, Man and Cybernetics, Part A. 39(3): 650–661.

Starke, P. H. 2003. INA: Integrated Net Analyzer. http://www2.informatik.huberlin.de /starke/ina.html.

Tricas, F. and J. Martinez. 1995. An extension of the liveness theory for concurrent sequential processes competing for shared resources, 3035–3040. *In* Proceedings of the IEEE International Conference on Systems, Man, and Cybernetics, Vancouver, British Columbia, Canada, Oct. 22–25.

Tricas, F., F. Garcia-Valles, J. M. Colom, and J. Ezpelata. 1998. A structural approach to the problem of deadlock prevention in processes with resources, 273–278. *In* Proceedings of WODES'98, Italy, 26–28 August.

Tricas, F., F. Garcia-Valles, J. M. Colom, and J. Ezpelata. An iterative method for deadlock prevention in FMS. pp. 139–148. *In* G. Stremersch [ed.]. 2000. Discrete Event Systems: Analysis and Control, Kluwer Academic, Boston MA.

Uzam, M. 2002. An optimal deadlock prevention policy for flexible manufacturing systems using Petri net models with resources and the theory of regions. International Journal of Advanced Manufacturing Technology. 19(3): 192–208.

Uzam, M. and M. C. Zhou. 2006. An improved iterative synthesis method for liveness enforcing supervisors of flexible manufacturing systems. International Journal of Production Research. 44(10): 1987–2030.

Uzam, M. and M. C. Zhou. 2007. An iterative synthesis approach to Petri net based deadlock prevention policy for flexible manufacturing systems. IEEE Transactions on Systems, Man and Cybernetics, Part A. 37(3): 362–371.

Xing, K. Y., B. S. Hu, and H. X. Chen. Deadlock avoidance policy for flexible manufacturing systems. pp. 239–263. *In* M. C. Zhou [ed.]. 1995. Petri Nets in Flexible and Agile Automation, Kluwer Academic, Boston MA.

Yamalidou, K. J., Moody, M. Lemmon, and P. Antsaklis. 1996. Feedback control of Petri nets based on place invariants. Automatica. 32(1): 15–28.

Zhou, M. C. and F. DiCesare. 1991. Parallel and sequential mutual exclusions for Petri net modeling of manufacturing systems with shared resources. IEEE Transactions on Robotics and Automation. 7(4): 515–527.

Chapter 6

Most Permissive Supervisors

ABSTRACT

This chapter presents a deadlock prevention approach to find a maximally permissive liveness-enforcing supervisor for an AMS if such a supervisor exists. Otherwise, it derives a most permissive one in the sense that there are no other pure Petri net supervisors that are more permissive than it. The presented approach computes the reachability graph of a plant model only once. A vector covering approach is used to reduce the sets of legal markings and FBMs to two smaller ones, namely, the minimal covering set of legal markings and the minimal covered set of FBMs, respectively. At each iteration, an FBM is selected and an ILPP is designed to compute a PI to forbid the FBM but no marking in the minimal covering set of legal markings. If the ILPP has no solution, implying that there is no maximally permissive pure Petri net supervisor for the plant net model, another ILPP is designed to remove the least number of legal markings whose reachability conditions contradict others. Then, a PI is redesigned to ensure the rest legal markings to be reachable. This process is carried out until no FBM can be reached. Finally, a most permissive liveness-enforcing supervisor is obtained. AMS examples are used to illustrate the reported approach.

6.1 INTRODUCTION

Behavioral permissiveness is an important criterion for the design of a liveness-enforcing Petri net supervisor. However, given a Petri net model, there may not exist a behaviorally optimal liveness-enforcing pure Petri net supervisor expressed by a set of control places (monitors). For example, the Petri net model of an AMS shown in Fig. 6.1 is an instance that cannot be optimally controlled by adding control places without self-loops. It has 39 reachable markings, 23 of which are legal, as shown in Fig. 6.2. For the optimal control purposes, an optimal supervisor should make the system live with all the 23 legal markings. However, it can be verified by the theory of regions in (Ghaffari et al., 2003) and (Uzam, 2002) that the net has 10 MTSIs and six of them have no monitor solutions, implying that there is no optimal pure Petri net supervisor that can lead to a live system with all the 23 legal markings.

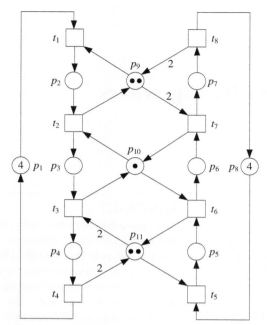

Figure 6.1 Petri net model of a manufacturing system.

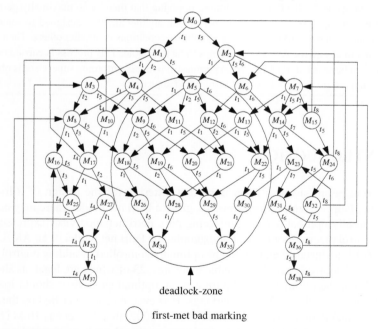

Figure 6.2 The reachability graph of the Petri net model in Fig. 6.1.

Now we discuss the nonexistence of an optimal pure net supervisor in detail. There are two legal markings $M_{L1} = (0\ 2\ 1\ 1\ 0\ 0\ 0\ 4\ 0\ 0\ 0)^T$ and $M_{L2} = (4\ 0\ 0\ 0\ 2\ 1\ 1\ 0\ 0\ 0\ 0)^T$, and a bad marking $M_B = (3\ 1\ 0\ 0\ 1\ 0\ 0\ 3\ 1\ 1\ 1)^T$ for this example. On the one hand, in order to prevent the net system from reaching deadlocks, M_B must be forbidden in the controlled system. On the other hand, for the optimal control purposes, both M_{L1} and M_{L2} should be reachable in the controlled system. It can be seen that both p_2 and p_5 have one token at M_B but they have two tokens at M_{L1} and M_{L2}, respectively. However, there does not exist a pure Petri net supervisor such that M_B is forbidden while M_{L1} and M_{L2} are reachable. In this case, it is appealing to design a most permissive liveness-enforcing supervisor, implying that the most legal markings are kept after adding control places and no other pure Petri net supervisors are more permissive than it.

Motivated by the work in (Chen *et al.*, 2011), this chapter presents an effective approach to obtain a most permissive liveness-enforcing supervisor for a Petri net model of an AMS if an optimal pure Petri net supervisor does not exist. The fact that no optimal control place can be designed is due to the presence of at least two legal markings that cannot be guaranteed to be reachable by forbidding a particular FBM. That is to say, when a PI (Lautenbach and Ridder, 1994; Yamalidou *et al.*, 1996) is designed to forbid an FBM, the reachability conditions of some legal markings mutually contradict. Thus, one or more legal markings are necessarily removed. This chapter deals with the problem by solving an ILPP such that the most legal markings are kept in the controlled system. In theory, this particular issue is motivated by the fact that not all finite automata admit a free-labeled Petri net representation. That is to say, an automaton supervisor does not always have a pure Petri net realization such that the behavior of the net is equivalent to the automaton (Li and Zhou, 2009).

The application scope of the presented method is the same as in (Chen *et al.*, 2011). That is to say, it can be used to design a most permissive liveness-enforcing supervisor for all AMS-oriented classes of Petri net models in the literature. A most permissive supervisor here means that most legal markings can be kept in the controlled net system. There are two cases for this problem: (i) If a net model has a maximally permissive pure Petri net supervisor, then the most permissive supervisor is maximally permissive. (ii) Otherwise, a most permissive supervisor is suboptimal but no other pure Petri net supervisors are more permissive than it.

6.2 BEST CONTROL PLACE SYNTHESIS

For the Petri net model of an AMS, there does not necessarily exist an optimal liveness-enforcing Petri net supervisor that is expressed by a set of control places. Then, deciding how to find a most permissive net supervisor such that no other supervisors are more permissive than it is interesting. This section aims to compute such a supervisor if the Petri net model of an AMS has no optimal supervisors represented by a set of control places without self-loops.

Definition 6.1 For an FBM $M \in \mathcal{M}^\star_{\text{FBM}}$, $\forall M_1, M_2 \in \mathcal{M}_L$, M_1 M-equals M_2 if $\forall p \in P_A$ with $M(p) \neq 0$, we have $M_1(p) = M_2(p)$, which is denoted as $M_1 =_M M_2$.

Definition 6.2 For an FBM $M \in \mathcal{M}^\star_{\text{FBM}}$, $\forall M_1 \in \mathcal{L}_M$, the set of legal markings that M-equal M_1 is defined as $\Xi_{M,M_1} = \{M_2 | M_2 \in \mathcal{M}_L, M_2 =_M M_1\}$.

Definition 6.2 indicates that when an FBM M is forbidden, if we cannot guarantee the reachability of legal marking M_1 in $\mathcal{L}_M$, then all legal markings that M-equal M_1 are forbidden. For example, in Fig. 6.2, $M_5 = (3\,1\,0\,0\,1\,0\,0\,3\,1\,1\,1)^T$ is an FBM and $M_{37} = (0\,2\,1\,1\,0\,0\,0\,4\,0\,0\,0)^T$ is legal. According to Definitions 6.1 and 6.2, we have $\Xi_{M_5,M_{37}} = \{M_3, M_{16}, M_{25}, M_{37}\}$ since $M_3(p_2) = M_{16}(p_2) = M_{25}(p_2) = M_{37}(p_2) = 2$ and $M_3(p_5) = M_{16}(p_5) = M_{25}(p_5) = M_{37}(p_5) = 0$. If a control place is designed to forbid M_5 and M_{37} must be forbidden as well, then all these four legal markings in $\Xi_{M_5,M_{37}}$ are forbidden.

Now we introduce a set of binary variables $f_1, f_2, \cdots,$ and $f_{N_{\mathcal{L}}}$ to the reachability condition for every legal marking in $\mathcal{L}_M$, where $N_{\mathcal{L}}$ is the number of elements in $\mathcal{L}_M$. Then, Eq. (5.13) is rewritten as:

$$\sum_{i \in \mathbb{N}_A} l_i \cdot (M_j(p_i) - M(p_i)) \leq Q \cdot f_j - 1, \quad \forall M_j \in \mathcal{L}_M \tag{6.1}$$

where Q is a positive constant integer that must be big enough and $f_j \in \{0, 1\}$. In Eq. (6.1), $f_j = 0$ indicates that the reachability condition of its corresponding marking M_j in $\mathcal{L}_M$ is satisfied and $f_j = 1$ indicates that this constraint will be redundant and the corresponding marking M_j in $\mathcal{L}_M$ is not ensured to be reachable. Let N_{M,M_j} be the total number of markings that M-equal M_j, i.e., $N_{M,M_j} = |\Xi_{M,M_j}|$. The following ILPP can be used to ensure the most legal markings to be reachable, which is denoted as the Most Legal Markings Problem (MLMP).

  MLMP:

$$\min \sum_{M_j \in \mathcal{L}_M} N_{M,M_j} \cdot f_j$$

  subject to
$$\sum_{i \in \mathbb{N}_A, M(p_i) \neq 0} l_i \cdot (M_j(p_i) - M(p_i)) \leq Q \cdot f_j - 1, \; \forall M_j \in \mathcal{L}_M$$
$$l_i \in \{0, 1, 2, \cdots\} \; \forall i \in \mathbb{N}_A \text{ and } M(p_i) \neq 0$$
$$f_j \in \{0, 1\} \; \forall M_j \in \mathcal{L}_M$$

The objective function represents the minimal number of legal markings that cannot be reached. Solving this ILPP, we have a solution about f_j. If $f_j = 1$ in the solution, it means that M_j cannot be reached if the control place is added. Thus, we should remove all the legal markings that M-equal M_j and compute a new set $\mathcal{L}_M$. A new ILPP is designed to ensure all the markings in the new set $\mathcal{L}_M$ to be reachable. Therefore, a most permissive supervisor can be obtained.

6.3 DEADLOCK PREVENTION POLICY

This section presents a deadlock prevention policy that can obtain a most permissive liveness-enforcing supervisor if an optimal one does not exist.

Algorithm 6.1 Most Permissive Deadlock Prevention Policy

Input: Petri net model (N, M_0) of an AMS with $N = (P^0 \cup P_A \cup P_R, T, F, W)$.

Output: An optimally controlled Petri net system (N_1, M_1) if it exists, or a most permissive controlled net system (N_1, M_1) that can reach as many legal markings as possible.

1) Compute the set of FBMs $\mathcal{M}_{FBM}$ and the set of legal markings $\mathcal{M}_L$ for (N, M_0).

2) Compute the minimal covered set of FBMs $\mathcal{M}_{FBM}^\star$ and the minimal covering set of legal markings $\mathcal{M}_L^\star$ for (N, M_0).

3) $V_M := \emptyset$. /* V_M is used to denote the set of control places to be computed.*/

4) **if** $\{\mathcal{M}_{FBM}^\star = \emptyset\}$ **then**

   $N_1 := N$, $M_1 := M_0$, go to Step 6.

   else

   $\forall M \in \mathcal{M}_{FBM}^\star$, compute the minimal covered set of M-related legal markings $\mathcal{L}_M$.

   Solve the set of integer linear inequalities of Eq. (5.13).

   while {the set of integer linear inequalities of Eq. (5.13) has no solution} **do**

      Solve an ILPP, i.e., MLMP, presented in Section 6.2.

      foreach $\{f_j = 1\}$ **do**

         $\mathcal{M}_L := \mathcal{M}_L - \varXi_{M,M_j}$.

      Compute the minimal covering set of legal markings $\mathcal{M}_L^\star$.

      Compute the minimal covering set of M-related legal markings $\mathcal{L}_M$.

   endwhile

   Let $l_i(i \in \mathbb{N}_A, M(p_i) \neq 0)$ be the solution.

   Design a PI I and a control place p_c by the method presented in Section 5.2.1.

   $V_M := V_M \cup \{p_c\}$, $\mathcal{M}_{FBM}^\star := \mathcal{M}_{FBM}^\star - F_I$. Denote the resulting net system as (N, M_0).

   endif

5) Go to Step 4.

6) Output (N_1, M_1).

7) End.

The above algorithm computes the reachability graph only once. We first use the vector covering approach. Then the sets of FBMs and legal markings are reduced to $\mathcal{M}_{FBM}^\star$ and $\mathcal{M}_L^\star$, respectively. This can reduce the number of inequalities for each PI and the total number of FBMs that should be forbidden by PIs. For each FBM M, the vector covering approach is used to reduce $\mathcal{M}_L^\star$ to $\mathcal{L}_M$. This step can further reduce the number of inequalities for each FBM. For each $M \in \mathcal{M}_{FBM}^\star$, if the ILPP has no solution, we introduce variables f_j's for all markings in $\mathcal{L}_M$ to represent their reachability. The objective function is minimized to ensure that the number of legal markings to be reached is as large as possible. If $f_j = 0$, that indicates that the corresponding bad marking M_j cannot be reached, we remove all the legal markings from $\mathcal{M}_L$ that are M-equal to M_j. A new set of legal markings is generated and the ILPP for the FBM is redesigned. If it has no solution, we design an MLMP to remove the minimal number of legal markings and redesign the ILPP. This process is carried out until the ILPP has a solution. As a result, a control place is designed to forbid the FBM and it can guarantee the reachability of as many legal markings as possible.

Now we discuss the computational complexity of the presented method. First, a complete enumeration of reachable markings is required, which is of exponential complexity with respect to the size of a Petri net. Second, we need solve an ILPP at each iteration, which is NP-hard. Therefore, the problem to be tackled in this chapter is NP-hard in theory. However, a vector covering approach is used to reduce the number of ILPPs and that of inequality constraints in each ILPP. An ILPP with a small number of inequalities can be solved efficiently. In (Chen *et al.*, 2011), we provide a large Petri net model with more than 80 million of legal markings while the largest number of inequalities in each ILPP is only 322 by using the vector covering approach. Thus, though the complexity of the considered problem is NP-hard in theory, it is still efficient from the experimental point of view.

If there exists a solution that satisfies Eq. (5.13) for each FBM in $\mathcal{M}^\star_{FBM}$, the presented control policy has the same results as the one in (Chen *et al.*, 2011). In this case, we have the following two theorems.

Theorem 6.1 *Algorithm 6.1 can obtain a maximally permissive supervisor for a Petri net model of an AMS if there exists a solution that satisfies Eq. (5.13) for each FBM in $\mathcal{M}^\star_{FBM}$.*

Proof Algorithm 6.1 does not terminate until all FBMs in $\mathcal{M}^\star_{FBM}$ are forbidden. According to Corollary 5.3, if all markings in $\mathcal{M}^\star_{FBM}$ are forbidden, then all FBMs are forbidden, implying that the controlled system is live since it cannot enter DZ any more. By Corollary 5.5, if Eq. (5.13) for each FBM has a solution, then every computed control place does not forbid any legal markings. That is to say, all legal markings are reachable and an optimal supervisor is obtained by Algorithm 6.1. □

Theorem 6.2 *Algorithm 6.1 can lead to a maximally permissive liveness-enforcing supervisor expressed by a set of control places if such a supervisor exists.*

Proof The proof of Theorem 6 in (Chen *et al.*, 2011) shows that if an optimal supervisor exists, there exists at least one solution that satisfies Eq. (5.13) for each FBM. According to Theorem 6.1, Algorithm 6.1 can obtain an optimal supervisor if there exists a solution that satisfies Eq. (5.13) for each FBM in $\mathcal{M}^\star_{FBM}$. Therefore, the conclusion holds. □

In the case that there is no optimal (maximally permissive) liveness-enforcing supervisor, we have the following result.

Theorem 6.3 *Algorithm 6.1 can obtain a most permissive liveness-enforcing supervisor, represented by a set of control places, for the Petri net model of an AMS if there does not exist an optimal Petri net one.*

Proof If an optimal pure Petri net supervisor does not exist, then there exists at least an FBM such that the set of integer linear inequalities of Eq. (5.13) has no solution. In this case, an ILPP MLMP is designed to remove the contradictory reachability conditions. From the objective function of the MLMP, it is known that the least legal markings are removed. This means that the control place can ensure as many legal markings as possible to be reachable. That is to say, the resulting supervisor is most permissive. □

In the following, two compact multiset formalisms are used to describe markings in $\mathcal{M}_L^\star$, $\mathcal{M}_{FBM}^\star$, and $\mathcal{L}_M$. The operation places are considered only for markings in $\mathcal{M}_L^\star$ and $\mathcal{M}_{FBM}^\star$. Thus, we use $\sum_{i \in \mathbb{N}_A} M(p_i)p_i$ to denote marking M in $\mathcal{M}_L^\star$ or $\mathcal{M}_{FBM}^\star$. For example, in Fig. 6.2, $M_{37} = (0\ 2\ 1\ 1\ 0\ 0\ 0\ 4\ 0\ 0\ 0)^T$ is a legal marking in $\mathcal{M}_L^\star$ and the operation place set is $P_A = \{p_2, p_3, p_4, p_5, p_6, p_7\}$. In this case, M_{37} is denoted by $M_{37} = 2p_2 + p_3 + p_4$. For an FBM $M \in \mathcal{L}_M$, the marked operation places at M are considered only. Thus, we use $\sum_{i \in \mathbb{N}_A, M(p_i) \neq 0} M'(p_i)p_i$ to denote marking M' in $\mathcal{L}_M$. In Fig. 6.2, $M_5 = (3\ 1\ 0\ 0\ 1\ 0\ 0\ 3\ 1\ 1\ 1)^T$ is an FBM and $M_{37} = (0\ 2\ 1\ 1\ 0\ 0\ 0\ 4\ 0\ 0\ 0)^T$ is a legal marking in $\mathcal{L}_{M_5}$. In this case, M_{37} is denoted by $M_{37} = 2p_2$ since operation places p_2 and p_5 are considered only to design a control place to forbid M_5.

The Petri net model of an AMS shown in Fig. 6.1 is used to illustrate the presented deadlock prevention policy. There are 11 places and eight transitions. The places can be classified into three groups: idle place set $P^0 = \{p_1, p_8\}$, operation place set $P_A = \{p_2 - p_7\}$, and resource place set $P_R = \{p_9, p_{10}, p_{11}\}$. The net has 39 reachable markings, 23 of which are legal and nine are FBMs, as shown in Fig. 6.2. The minimal covered set of FBMs $\mathcal{M}_{FBM}^\star$ and the minimal covering set of legal markings $\mathcal{M}_L^\star$ have three and two elements, respectively. Specifically, we have $\mathcal{M}_{FBM}^\star = \{p_2 + p_5, p_3 + p_5, p_2 + p_6\}$ and $\mathcal{M}_L^\star = \{2p_5 + p_6 + p_7, 2p_2 + p_3 + p_4\}$.

At the first iteration, $FBM_1 = p_2 + p_5$ is selected. We have $\mathcal{L}_{FBM_1} = \{2p_5, 2p_2\}$. In order to forbid FBM_1, I_1 is designed to satisfy constraint $l_2 \cdot \mu_2 + l_5 \cdot \mu_5 \leq l_2 \cdot 1 + l_5 \cdot 1 - 1$. For the optimal control purposes, I_1 must not forbid any markings in $\mathcal{L}_{FBM_1}$. Thus, I_1 has to satisfy Eq. (5.13):

$$-l_2 + l_5 \leq -1 \quad \text{and}$$
$$l_2 - l_5 \leq -1.$$

The above integer linear program has no solution. Thus, we introduce a set of variables f_j's ($j = 1, 2$) to represent the reachability of the two legal markings in $\mathcal{L}_{FBM_1}$. Let $M_{L1} = 2p_5$ and $M_{L2} = 2p_2$. The number of legal markings that FBM_1-equal M_{L1} is four, i.e., $|\Xi_{FBM_1, M_{L1}}| = 4$. Similarly, we have $|\Xi_{FBM_1, M_{L2}}| = 4$. Now, MLMP is presented as follows:

MLMP:

$$\text{min} \quad 4 \cdot f_1 + 4 \cdot f_2$$
$$\text{subject to}$$
$$-l_2 + l_5 \leq Q \cdot f_1 - 1$$
$$l_2 - l_5 \leq Q \cdot f_2 - 1$$
$$l_2, l_5 \in \{1, 2, \cdots\}$$
$$f_1, f_2 \in \{0, 1\}$$

The above MLMP has a solution with $f_1 = 1$ and $f_2 = 0$. Variable $f_1 = 1$ means that M_{L1} cannot be reached for the most permissive control purposes. Thus, all the markings that FBM$_1$-equal M_{L1} should be removed. After removing the four markings in $\varXi_{\text{FBM}_1, M_{L1}}$, the resulting set of legal markings has 19 elements. By using the vector covering approach, $\mathcal{M}_L^\star$ has two markings, i.e., $\mathcal{M}_L^\star = \{p_5 + p_6 + p_7, 2p_2 + p_3 + p_4\}$. Therefore, we have $\mathcal{L}_{\text{FBM}_1} = \{p_5, 2p_2\}$. In order to forbid FBM$_1$, I_1 is redesigned to satisfy constraint $l_2 \cdot \mu_2 + l_5 \cdot \mu_5 \leq l_2 \cdot 1 + l_5 \cdot 1 - 1$. For the most permissive control purposes, I_1 must not forbid any markings in $\mathcal{L}_{\text{FBM}_1}$. Thus, I_1 has to satisfy Eq. (5.13):

$$-l_2 \leq -1 \quad \text{and}$$
$$l_2 - l_5 \leq -1.$$

The above integer linear program has a solution with $l_2 = 1$ and $l_5 = 2$. Then, a control place p_{c_1} can be designed for I_1: $\mu_2 + 2\mu_5 + \mu_{p_{c_1}} = 2$ by the method presented in Section 5.2.1. As a result, we have $M_0(p_{c_1}) = 2$, $^\bullet p_{c_1} = \{t_2, 2t_6\}$, and $p_{c_1}^\bullet = \{t_1, 2t_5\}$. Only FBM$_1$ is forbidden by I_1, i.e., $F_{I_1} = \{\text{FBM}_1\}$. Hence, we have $\mathcal{M}_{\text{FBM}}^\star := \mathcal{M}_{\text{FBM}}^\star - \{\text{FBM}_1\}$, i.e., $\mathcal{M}_{\text{FBM}}^\star = \{p_3 + p_5, p_2 + p_6\}$.

At the second iteration, FBM$_2 = p_3 + p_5$ is selected. We have $\mathcal{L}_{\text{FBM}_2} = \{p_5, p_3\}$. In order to forbid FBM$_2$, I_2 is designed to satisfy constraint: $l_3 \cdot \mu_3 + l_5 \cdot \mu_5 \leq l_3 \cdot 1 + l_5 \cdot 1 - 1$. For the most permissive control purposes, I_2 must not forbid any markings in $\mathcal{L}_{\text{FBM}_2}$. Thus, I_2 has to satisfy Eq. (5.13):

$$-l_3 \leq -1 \quad \text{and}$$
$$-l_5 \leq -1.$$

The above integer linear system has a solution with $l_3 = 1$ and $l_5 = 1$. Then, a control place p_{c_2} can be designed for I_2: $\mu_3 + \mu_5 + \mu_{p_{c_2}} = 1$ by the method presented in Section 5.2.1. As a result, we have $M_0(p_{c_2}) = 1$, $^\bullet p_{c_2} = \{t_3, t_6\}$, and $p_{c_2}^\bullet = \{t_2, t_5\}$. Only FBM$_2$ is forbidden by I_2, i.e., $F_{I_2} = \{\text{FBM}_2\}$. Hence, we have $\mathcal{M}_{\text{FBM}}^\star := \mathcal{M}_{\text{FBM}}^\star - \{\text{FBM}_2\}$, i.e., $\mathcal{M}_{\text{FBM}}^\star = \{p_2 + p_6\}$.

At the last iteration, FBM$_3 = p_2 + p_6$ is selected. We have $\mathcal{L}_{\text{FBM}_3} = \{p_6, 2p_2\}$. In order to forbid FBM$_3$, I_3 is designed to satisfy constraint $l_2 \cdot \mu_2 + l_6 \cdot \mu_6 \leq l_2 \cdot 1 + l_6 \cdot 1 - 1$. For the most permissive control purposes, I_3 must not forbid any markings in $\mathcal{L}_{\text{FBM}_3}$. Thus, I_3 also has to satisfy Eq. (5.13):

$$-l_2 \leq -1 \quad \text{and}$$
$$l_2 - l_6 \leq -1.$$

The above integer linear program has a solution with $l_2 = 1$ and $l_6 = 2$. Then, a control place p_{c_3} can be designed for I_3: $\mu_2 + 2\mu_6 + \mu_{p_{c_3}} = 2$. As a result, we have

$M_0(p_{c_3}) = 2$, $^\bullet p_{c_3} = \{t_2, 2t_7\}$, and $p_{c_3}^\bullet = \{t_1, 2t_6\}$. Only FBM$_3$ is forbidden by I_3, i.e., $F_{I_3} = \{FBM_3\}$. Hence, we have $\mathcal{M}_{FBM}^\star := \mathcal{M}_{FBM}^\star - \{FBM_3\}$, i.e., $\mathcal{M}_{FBM}^\star = \emptyset$.

Now, Algorithm 6.1 terminates and three control places are obtained. Table 6.1 shows its outcomes, where the first column is the iteration number i, the second represents the selected FBM (FBM$_i$), the third, N_{LP}, is the number of inequalities to determine the PI for FBM$_i$, and the fourth shows the PI (I_i) for FBM$_i$. The fifth to seventh columns indicate pre-transitions ($^\bullet p_{c_i}$), post-transitions ($p_{c_i}^\bullet$), and initial marking ($M_0(p_{c_i})$) of control place p_{c_i}, respectively. Adding the three control places to the original net model, the resulting net is live with 19 legal markings. This represents the most legal behavior that we can have by adding control places. That is to say, no other pure Petri net supervisors can be found, which can lead to a live controlled system that has more than 19 reachable states.

Table 6.1 Control places computed for the Petri net model shown in Fig. 6.1

i	FBM$_i$	N_{LP}	I_i	$^\bullet p_{c_i}$	$p_{c_i}^\bullet$	$M_0(p_{c_i})$
1	$p_2 + p_5$	2	$\mu_2 + 2\mu_5 \leq 2$	$t_2, 2t_6$	$t_1, 2t_5$	2
2	$p_3 + p_5$	2	$\mu_3 + \mu_5 \leq 1$	t_3, t_6	t_2, t_5	1
3	$p_2 + p_6$	2	$\mu_2 + 2\mu_6 \leq 2$	$t_2, 2t_7$	$t_1, 2t_6$	2

According to this example, it can be seen that the contradictory reachability conditions are removed if there is no optimal control place for an FBM. By using an integer linear programming approach, as many legal markings as possible are kept. In this process, since the vector covering approach is used to reduce the set of legal markings, the number of inequalities is much less than that of legal markings. This can obviously alleviate the computational burden.

6.4 EXPERIMENTAL RESULTS

A Petri net model is shown in Fig. 6.3, which is an S^{4}PR from (Tricas *et al.*, 2005). There are nine places and seven transitions. It has the following place partitions: $P^0 = \{p_{10}, p_{20}\}$, $P_R = \{p_{31}, p_{32}\}$, and $P_A = \{p_{11}, p_{12}, p_{13}, p_{21}, p_{22}\}$. It has 20 reachable markings, 6 and 13 of which are FBMs and legal ones, respectively. Using the vector covering approach, both $\mathcal{M}_{FBM}^\star$ and $\mathcal{M}_L^\star$ have three markings only, i.e., $\mathcal{M}_{FBM}^\star = \{2p_{12}, p_{12} + p_{21}, p_{11} + p_{21}\}$ and $\mathcal{M}_L^\star = \{p_{11} + p_{12}, 2p_{21} + p_{22}, 2p_{11} + p_{13}\}$. Table 6.2 shows the application of the presented policy. Note that at the third iteration, control place p_{c_3} is suboptimal since there is no an optimal solution for FBM$_3 = p_{11} + p_{21}$. In this case, we design MLMP to obtain a most permissive one as denoted by p_{c_3} for PI: $\mu_{11} + 2\mu_{21} \leq 2$, as shown in the last row in Table 6.2. Finally, there are totally three control places to be added and the resulting net is live with 11 reachable markings that represent the most permissive behavior.

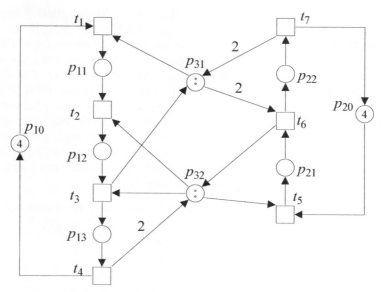

Figure 6.3 A Petri net model in (Tricas *et al.*, 2005).

Table 6.2 Control places for the net model in Fig. 6.3

i	FBM_i	N_{LP}	I_i	${}^\bullet p_{c_i}$	$p_{c_i}^\bullet$	$M_0(p_{c_i})$
1	$2p_{12}$	1	$\mu_{12} \leq 1$	t_3	t_2	1
2	$p_{12} + p_{21}$	2	$2\mu_{12} + \mu_{21} \leq 2$	$2t_3, t_6$	$2t_2, t_5$	2
3	$p_{11} + p_{21}$	2	$\mu_{11} + 2\mu_{21} \leq 2$	$t_2, 2t_6$	$t_1, 2t_5$	2

In (Tricas *et al.*, 2005), Tricas *et al.* add four control places to this model and the controlled model has 10 reachable markings. However, considering the presented solution to this problem, we obtain a most permissive liveness-enforcing supervisor with less control places and the controlled system has more permissive behavior. Table 6.3 shows the performance comparison of the two policies for this example.

Table 6.3 Performance comparison of typical deadlock control policies for the S^4PR in Fig. 6.3

Parameters	(Tricas *et al.*, 2005)	Alg. 6.1
No. monitors	4	3
No. states	10	11

Consider another Petri net model of an AMS, as shown in Fig. 6.4. It is a WS^3PR that is a weighted version of an S^3PR in (Ezpeleta *et al.*, 1995). Comparing with S^3PR, both operation places p_{13} and p_{18} in this model have weighted resource requirements. There are 26 places and 20 transitions. It has the following place partitions: $P^0 = \{p_1, p_5, p_{14}\}$, $P_R = \{p_{20}, \ldots, p_{26}\}$, and $P_A = \{p_2, \ldots, p_4, p_6, \ldots, p_{13}, p_{15}, \ldots, p_{19}\}$. It has 9,572 reachable markings, 1,762 and 5,395 of which are FBMs and legal ones, respectively. Using the vector covering approach, $\mathcal{M}_{FBM}^{\star}$ and $\mathcal{M}_{L}^{\star}$ have 15 and 87 markings, respectively. Table 6.4 shows the application of the presented policy. This net does not admit an optimal pure Petri net supervisor since at the first iteration, there is no optimal control place for $FBM_1 = p_{11} + p_{16}$. Thus, p_{c_1} is not optimal and its addition prevents 600 legal markings from being reached. For all the other control places, there are no legal markings lost after their addition. Finally, there are 12 control places to be added and the controlled net is live with 4,795 reachable markings. This is the most legal behavior by adding control places to the original Petri net model.

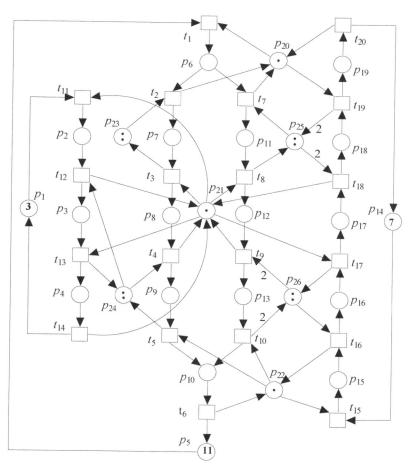

Figure 6.4 WS³PR model of an AMS.

Table 6.4 Control places computed for the net model in Fig. 6.4

i	FBM$_i$	N_{LP}	I_i	$^\bullet p_{c_i}$	$p_{c_i}^\bullet$	$M_0(p_{c_i})$
1	$p_{11}+p_{16}$	2	$2\mu_{11}+\mu_{16}\le 2$	$2t_8,t_{17}$	$2t_7,t_{16}$	2
2	$p_{11}+p_{17}$	2	$\mu_{11}+\mu_{17}\le 1$	t_8,t_{18}	t_7,t_{17}	1
3	$p_6+2p_7+p_8+p_9+$ $p_{15}+2p_{16}+p_{18}$	10	$\mu_6+\mu_7+\mu_8+\mu_9+$ $2\mu_{15}+2\mu_{16}+\mu_{18}\le 11$	$t_5,t_7,2t_{17},t_{19}$	$t_1,2t_{15},t_{18}$	11
4	$p_2+p_3+p_9+p_{15}+$ $2p_{16}$	9	$\mu_2+\mu_3+\mu_9+\mu_{15}+$ $\mu_{16}\le 5$	t_5,t_{13},t_{17}	t_4,t_{11},t_{15}	5
5	$p_3+p_8+p_9+p_{15}+$ $2p_{16}$	9	$\mu_3+\mu_8+\mu_9+\mu_{15}+$ $\mu_{16}\le 5$	t_5,t_{13},t_{17}	t_3,t_{12},t_{15}	5
6	$p_6+2p_7+2p_9+$ $p_{15}+2p_{16}+p_{17}$	6	$\mu_6+\mu_7+\mu_9+\mu_{15}+$ $\mu_{16}+\mu_{17}\le 8$	t_3,t_5,t_7,t_{18}	t_1,t_4,t_{15}	8
7	$p_6+2p_7+p_{17}+p_{18}$	4	$\mu_6+\mu_7+\mu_{17}+\mu_{18}\le 4$	t_3,t_7,t_{19}	t_1,t_{17}	4
8	p_2+2p_3	2	$\mu_2+\mu_3\le 2$	t_{13}	t_{11}	2
9	$2p_3+p_8$	2	$\mu_3+\mu_8\le 2$	t_4,t_{13}	t_3,t_{12}	2
10	$p_{13}+p_{15}$	2	$\mu_{13}+\mu_{15}\le 1$	t_{10},t_{16}	t_9,t_{15}	1
11	$p_{11}+p_{15}$	2	$\mu_{11}+\mu_{15}\le 1$	t_8,t_{16}	t_7,t_{15}	1
12	$p_{12}+p_{15}$	2	$\mu_{12}+\mu_{15}\le 1$	t_9,t_{16}	t_8,t_{15}	1

6.5 CONCLUSIONS

This chapter presents a deadlock prevention policy to derive an optimal, i.e., maximally permissive, pure Petri net supervisor if such a supervisor exists. Otherwise, a most permissive supervisor is obtained, which keeps as many legal markings as possible. Behavioral permissiveness is very important in designing a liveness-enforcing supervisor for a system to be controlled. An optimal liveness-enforcing supervisor can lead to high utilization of system resources. The theory of regions (Ghaffari et al., 2003) can always derive an optimal supervisor if such a supervisor exists. However, it suffers from the computational complexity problem. In (Chen et al., 2011), a novel method to obtain an optimal liveness-enforcing supervisor is developed by using a vector covering approach. However, there is no existing method to obtain a most permissive liveness-enforcing supervisor if an optimal one does not exist. The presented method can be applied to all classes of Petri net models of AMSs, which are the same as Uzam and Zhou's method (Uzam and Zhou, 2006) that also suffers from the state explosion problem since it requires a complete enumeration of reachable markings. The vector covering approach is used to greatly reduce the number of sets of inequalities and that of inequalities in each set. Thus, the computational costs are drastically reduced.

We do not provide an objective function in the integer linear program for each FBM since any feasible solution is acceptable. It is possible to add an objective function to reduce the structural complexity of each control place, which can lead to a structurally simpler supervisor, thereby reducing the implementation cost (Hu et al., 2010).

6.6 BIBLIOGRAPHICAL REMARKS

Maximal permissiveness is an important criterion to evaluate the performance of a liveness-enforcing Petri net supervisor. However, only a few methods can lead to an optimal, i.e., maximally permissive, Petri net supervisor. The approaches that can definitely find an optimal supervisor for a Petri net can be found in (Chen *et al.*, 2011, 2012b; Chen and Li, 2011; Ghaffari *et al.*, 2003; Uzam, 2002). Also, an efficient optimal deadlock prevention policy that focuses on the subclasses of Petri nets can be found in (Li *et al.*, 2008). Piroddi *et al.* (2008, 2009) provide effective methods that can find Petri net supervisors with very high permissiveness for PT-ordinary nets. In fact, they find an optimal supervisor for every example in the two papers (Piroddi *et al.*, 2008, 2009) but cannot prove the optimality of their methods in theory. A lot of work develops suboptimal supervisors such as (Ezpeleta *et al.*, 1995; Huang *et al.*, 2001; Li *et al.*, 2004, 2007; Li and Zhou, 2006). Most materials reported in this chapter for the design of most permissive supervisors come from a recent work (Chen *et al.*, 2012a).

References

Chen, Y. F. and Z. W. Li. 2011. Design of a maximally permissive liveness-enforcing supervisor with a compressed supervisory structure for flexible manufacturing systems. Automatica. 47(5): 1028–1034.

Chen, Y. F., Z. W. Li, M. Khalgui, and O. Mosbahi. 2011. Design of a maximally permissive liveness-enforcing Petri net supervisor for flexible manufacturing systems. IEEE Transactions on Automation Science and Engineering. 8(2): 474–493.

Chen, Y. F., Z. W. Li, and M. C. Zhou. 2012a. Most permissive liveness-enforcing Petri net supervisors for flexible manufacturing systems. International Journal of Production Research. 50(22): 6357–6371.

Chen, Y. F., Z. W. Li, and M. C. Zhou. 2012b. Behaviorally optimal and structurally simple liveness-enforcing supervisors of flexible manufacturing systems. IEEE Transactions on Systems, Man, and Cybernetics, Part A. 42(3): 615–629.

Ezpeleta, J., J. M. Colom, and J. Martinez. 1995. A Petri net based deadlock prevention policy for flexible manufacturing systems. IEEE Transactions on Robotics and Automation. 11(2): 173–184.

Ghaffari, A., N. Rezg, and X. L. Xie. 2003. Design of a live and maximally permissive Petri net controller using the theory of regions. IEEE Transactions on Robotics and Automation. 19(1): 137–142.

Hu, H. S., M. C. Zhou, and Z. W. Li. 2010. Low-cost high-performance supervision in ratio-enforced automated manufacturing systems using timed Petri nets. IEEE Transactions on Automation Science and Engineering. 7(4): 933–944.

Huang, Y. S., M. D. Jeng, X. L. Xie, and S. L. Chung. 2001. Deadlock prevention based on Petri nets and siphons. International Journal of Production Research. 39(2): 283–305.

Lautenbach, K. and H. Ridder. 1994. Liveness in bounded Petri nets which are covered by T-invariants. Lecture Notes in Computer Science. 815: 358–375.

Li, Z. W. and M. C. Zhou. 2004. Elementary siphons of Petri nets and their application to deadlock prevention in flexible manufacturing systems. IEEE Transactions on Systems, Man, and Cybernetics, Part A. 34(1): 38–51.

Li, Z. W., H. S. Hu, and A. R. Wang. 2007. Design of liveness-enforcing supervisors for flexible manufacturing systems using Petri nets. IEEE Transactions on Systems, Man, and Cybernetics, Part C. 37(4): 517–526.

Li, Z. W. and M. C. Zhou. 2006. Two-stage method for synthesizing liveness-enforcing supervisors for flexible manufacturing systems using Petri nets. IEEE Transactions on Industrial Informatics. 2(4): 313–325.

Li, Z. W. and M. C. Zhou. 2009. Deadlock Resolution in Automated Manufacturing Systems: A Novel Petri Net Approach. Berlin: Springer-Verlag.

Li, Z. W., M. C. Zhou, and M. D. Jeng. 2008. A maximally permissive deadlock prevention policy for FMS based on Petri net siphon control and the theory of regions. IEEE Transactions on Automation Science and Engineering. 5(1): 182–188.

Piroddi, L., R. Cordone, and I. Fumagalli. 2008. Selective siphon control for deadlock prevention in Petri nets. IEEE Transactions on Systems, Man, and Cybernetics, Part A. 38(6): 1337–1348.

Piroddi, L., R. Cordone, and I. Fumagalli. 2009. Combined siphon and marking generation for deadlock prevention in Petri nets. IEEE Transactions on Systems, Man and Cybernetics, Part A. 39(3): 650–661.

Tricas, F., F. Garcia-Valles, J. M. Colom, and J. Ezpelata. 2005. A Petri net structure-based deadlock prevention solution for sequential resource allocation systems, 271–277. *In* The Proceedings of the IEEE Transactions Conference on Robotics and Automation, Apr. 18–22, Barcelona, Spain.

Uzam, M., 2002. An optimal deadlock prevention policy for flexible manufacturing systems using Petri net models with resources and the theory of regions. International Journal of Advanced Manufacturing Technology. 19(3): 192–208.

Uzam, M. and M. C. Zhou. 2006. An improved iterative synthesis method for liveness enforcing supervisors of flexible manufacturing systems. International Journal of Production Research. 44(10): 1987–2030.

Yamalidou, K., J. Moody, M. Lemmon, and P. Antsaklis. 1996. Feedback control of Petri nets based on place invariants. Automatica. 32(1): 15–28.

Chapter 7

Structurally Minimal Supervisors

ABSTRACT

In this chapter, an optimal deadlock prevention policy for AMSs is presented, which can obtain a maximally permissive liveness-enforcing supervisor while the number of control places is minimized. By using the vector covering approach, the sets of legal markings and FBMs are reduced to two small ones, i.e., the minimal covering set of legal markings and the minimal covered set of FBMs. An optimal control purpose can be achieved by designing control places such that all markings in the minimal covered set of FBMs are not reachable and no marking in the minimal covering set of legal markings is forbidden. There is no need to add a control place for each FBM since multiple FBMs can be forbidden by only one optimal control place. In this case, we design an ILPP to minimize the number of control places to be computed. The resulting net has the minimal number of control places and is maximally permissive. A number of AMS examples from the literature are used to illustrate the presented method.

7.1 INTRODUCTION

Petri nets (Murata, 1989) are widely employed to model and control AMSs since they are suitable to detect deadlocks of a system and develop a policy to deal with them (Ezpeleta *et al.*, 1995; Ghaffari *et al.*, 2003; Huang *et al.*, 2001; Li and Zhou, 2009; Reveliotis, 2007). Deadlock prevention is one of the most important approaches whose goal is to impose constraints on a system to prevent it from reaching deadlock states (Ezpeleta *et al.*, 1995; Fanti and Zhou, 2004, 2005; Jeng and Xie, 2005; Piroddi *et al.*, 2008, 2009). In this case, the computation is carried out off-line and once the control policy is established and applied, deadlocks can no longer occur.

Generally, we consider three important criteria in evaluating and designing a liveness-enforcing supervisor for a system to be controlled: behavioral permissiveness, structural complexity, and computational complexity. Chapter 5 shows a method that can definitely derive a maximally permissive liveness-enforcing supervisor for Petri net models of AMSs if such a supervisor exists, where a maximally permissive control place is designed by a PI (Lautenbach and Ridder, 1994; Yamalidou *et al.*, 1996) that forbids one of the FBMs and no legal marking is prohibited. The PI is computed by solving an ILPP. In order to make the considered problem tractable, a vector covering approach is developed to reduce the sets of legal markings and FBMs to be small, i.e., the minimal covering set of legal markings and the minimal covered set of FBMs. Then, only two compressed sets are considered in the design of a supervisor. In this case, the computational burden is greatly reduced. However, the previous work does not ensure that the supervisor has a compact structure and hence suffers from the structural complexity problem in theory.

This chapter focuses on the optimization of the two criteria: behavioral permissiveness and structural complexity. A non-iterative approach is presented, aiming to overcome the structural complexity problem and ensure that the controlled system is still maximally permissive. First, the vector covering approach is used to lessen the sets of legal markings and FBMs to be small. Then, an ILPP is designed to satisfy the following three conditions:

1. Each FBM in the minimal covered set of FBMs is forbidden by at least one control place. Therefore, all the FBMs are forbidden, implying that the controlled system is live.
2. No markings in the minimal covering set of legal markings are forbidden. Therefore, no legal markings are excluded, implying that the final supervisor is maximally permissive.
3. The objective function minimizes the number of control places to be added under an assumption that a control place is associated with a P-semiflow.

If the ILPP has a solution, we can obtain a maximally permissive supervisor with a minimal structure, where each control place is associated with a P-semiflow. Then, the controlled system can be implemented with less hardware and software costs, and the resources of the system are highly utilized as well. This method has the same application scope as the one in Chapter 5, i.e., all classes of Petri net models of AMSs including PPN (Hsieh and Chang, 1994; Xing *et al.*, 1995), S^3PR (Ezpeleta *et al.*, 1995), ES^{3}PR (Tricas *et al.*, 1998), S^4PR (Tricas *et al.*, 2000), S^*PR (Ezpeleta *et al.*, 2002), S^2LSPR (Park and Reveliotis, 2000), S^3PGR^2 (Park and Reveliotis, 2001), and S^3PMR (Huang *et al.*, 2006). Though it is both behaviorally and structurally optimal, it still suffers from the computational complexity problem that makes it inapplicable to large-scale Petri net models. Also, it cannot be applied to Petri net instances in the above models that have no maximally permissive liveness-enforcing supervisor expressed by control places without self-loops.

7.2 SYNTHESIS OF MINIMUM CONTROL PLACES

This section presents an approach to minimize the number of control places to be computed. In the following, we define $\mathbb{N}^{\star}_{\text{FBM}}$ as $\mathbb{N}^{\star}_{\text{FBM}} = \{i | M_i \in \mathcal{M}^{\star}_{\text{FBM}}\}$.

A PI associated with a control place may forbid more than one FBM. Given I_j for an FBM $M_j \in \mathcal{M}^{\star}_{\text{FBM}}$, any FBM $M_k \in \mathcal{M}^{\star}_{\text{FBM}}$ ($k \neq j$) is forbidden if it satisfies the following constraint:

$$\sum_{i\in\mathbb{N}_A} l_{j,i}\cdot M_k(p_i) \geq \sum_{i\in\mathbb{N}_A} l_{j,i}\cdot M_j(p_i),\ \forall M_k \in \mathcal{M}_{\text{FBM}}^{\star}\ \text{and}\ k\neq j \qquad (7.1)$$

where $l_{j,i}$'s $(i\in\mathbb{N}_A)$ are the coefficients of I_j. By simplifying Eq. (7.1), we have

$$\sum_{i\in\mathbb{N}_A} l_{j,i}\cdot(M_k(p_i)-M_j(p_i)) \geq 0,\ \forall M_k \in \mathcal{M}_{\text{FBM}}^{\star}\ \text{and}\ k\neq j \qquad (7.2)$$

For I_j, we introduce a set of variables $f_{j,k}$'s ($k\in\mathbb{N}_{\text{FBM}}^{\star}$ and $k\neq j$) to represent the relation between I_j and any marking M_k in $\mathcal{M}_{\text{FBM}}^{\star}$. Then, for each FBM in $\mathcal{M}_{\text{FBM}}^{\star}$, Eq. (7.2) is rewritten as:

$$\sum_{i\in\mathbb{N}_A} l_{j,i}\cdot(M_k(p_i)-M_j(p_i)) \geq -Q\cdot(1-f_{j,k}),\ \forall M_k \in \mathcal{M}_{\text{FBM}}^{\star}\ \text{and}\ k\neq j \qquad (7.3)$$

where Q is a positive constant integer that must be big enough and $f_{j,k}\in\{0,1\}$. In Eq. (7.3), $f_{j,k}=1$ indicates that M_k is forbidden by I_j and $f_{j,k}=0$ indicates that this constraint is redundant and M_k cannot be forbidden by I_j.

Now we introduce another set of variables h_j's ($j\in\mathbb{N}_{\text{FBM}}^{\star}$) for I_j, where $h_j=1$ represents that I_j is selected to compute a control place and $h_j=0$ indicates that I_j is redundant and it is not necessary to add a control place for it. If I_j is not selected, it cannot forbid any FBM. Therefore, we have the following condition:

$$f_{j,k} \leq h_j,\ \forall k\in\mathbb{N}_{\text{FBM}}^{\star}\ \text{and}\ k\neq j \qquad (7.4)$$

Any FBM M_j in $\mathcal{M}_{\text{FBM}}^{\star}$ must be forbidden by at least one PI. Thus, we have the following constraint:

$$h_j + \sum_{k\in\mathbb{N}_{\text{FBM}}^{\star},\ k\neq j} f_{k,j} \geq 1 \qquad (7.5)$$

For every FBM in $\mathcal{M}_{\text{FBM}}^{\star}$, we have Eqs.(7.3), (7.4), and (7.5). Combining all these constraints and the reachability conditions together, we have the following ILPP that is denoted as the Minimal number of Control Places Problem (MCPP).

MCPP:

$$\min \sum_{j\in\mathbb{N}_{\text{FBM}}^{\star}} h_j$$

subject to

$$\sum_{i\in\mathbb{N}_A} l_{j,i}\cdot(M_l(p_i)-M_j(p_i)) \leq -1,\ \forall M_j \in \mathcal{M}_{\text{FBM}}^{\star}\ \text{and}\ \forall M_l \in \mathcal{M}_L^{\star} \qquad (7.6)$$

$$\sum_{i\in\mathbb{N}_A} l_{j,i}\cdot(M_k(p_i)-M_j(p_i)) \geq -Q\cdot(1-f_{j,k}),\ \forall M_j, M_k \in \mathcal{M}_{\text{FBM}}^{\star}\ \text{and}\ j\neq k \qquad (7.7)$$

$$f_{j,k} \leq h_j, \ \forall j,k \in \mathbb{N}^\star_{\text{FBM}} \text{ and } j \neq k \tag{7.8}$$

$$h_j + \sum_{k \in \mathbb{N}^\star_{\text{FBM}}, \ k \neq j} f_{k,j} \geq 1, \ \forall j \in \mathbb{N}^\star_{\text{FBM}} \tag{7.9}$$

$$l_{j,i} \in \{0,1,2,\cdots\}, \ \forall i \in \mathbb{N}_A \text{ and } \forall j \in \mathbb{N}^\star_{\text{FBM}}$$

$$f_{j,k} \in \{0,1\}, \ \forall j,k \in \mathbb{N}^\star_{\text{FBM}} \text{ and } j \neq k$$

$$h_j \in \{0,1\}, \ \forall j \in \mathbb{N}^\star_{\text{FBM}}$$

The objective function represents the minimal number of control places to be computed. We discuss MCPP in detail with respect to the numbers of constraints and variables. Let us consider the number of constraints of each type. The type of Eq. (7.6) has $|\mathcal{M}^\star_{\text{FBM}}| \cdot |\mathcal{M}^\star_L|$ constraints. Since $M_k \neq M_j$, the number of constraints of type Eq. (7.7) is $|\mathcal{M}^\star_{\text{FBM}}| \cdot (|\mathcal{M}^\star_{\text{FBM}}| - 1)$. Similarly, the type of Eq. (7.8) has $|\mathcal{M}^\star_{\text{FBM}}| \cdot (|\mathcal{M}^\star_{\text{FBM}}| - 1)$ constraints. Finally, the number of constraints of type Eq. (7.9) is $|\mathcal{M}^\star_{\text{FBM}}|$. The total number of all the four types of constraints in MCPP is $|\mathcal{M}^\star_{\text{FBM}}| \cdot (|\mathcal{M}^\star_L| + 2|\mathcal{M}^\star_{\text{FBM}}| - 1)$. Now we consider the number of variables in MCPP. Let $|P_A|$ be the number of operation places in a Petri net model to be controlled. The number of variables $l_{j,i}$'s ($i \in \mathbb{N}_A, j \in \mathbb{N}^\star_{\text{FBM}}$) is $|P_A| \cdot |\mathcal{M}^\star_{\text{FBM}}|$. The number of variables $f_{j,k}$'s is $|\mathcal{M}^\star_{\text{FBM}}| \cdot (|\mathcal{M}^\star_{\text{FBM}}| - 1)$ since $j,k \in \mathbb{N}^\star_{\text{FBM}}$ and $j \neq k$. The number of variables h_j's ($j \in \mathbb{N}^\star_{\text{FBM}}$) is equal to that of markings in $\mathcal{M}^\star_{\text{FBM}}$, i.e., $|\mathcal{M}^\star_{\text{FBM}}|$. Therefore, MCPP has $|\mathcal{M}^\star_{\text{FBM}}| \cdot (|P_A| + |\mathcal{M}^\star_{\text{FBM}}|)$ variables in total. Table 7.1 summarizes the information of MCPP.

Table 7.1 The numbers of constraints and variables in MCPP

Constraint type	No. of constraints	Variable name	No. of variables
(7.6)	$|\mathcal{M}^\star_{\text{FBM}}| \cdot |\mathcal{M}^\star_L|$	$l_{j,i}$	$|\mathcal{M}^\star_{\text{FBM}}| \cdot |P_A|$
(7.7)	$|\mathcal{M}^\star_{\text{FBM}}| \cdot (|\mathcal{M}^\star_{\text{FBM}}| - 1)$	$f_{j,k}$	$|\mathcal{M}^\star_{\text{FBM}}| \cdot (|\mathcal{M}^\star_{\text{FBM}}| - 1)$
(7.8)	$|\mathcal{M}^\star_{\text{FBM}}| \cdot (|\mathcal{M}^\star_{\text{FBM}}| - 1)$	h_j	$|\mathcal{M}^\star_{\text{FBM}}|$
(7.9)	$|\mathcal{M}^\star_{\text{FBM}}|$		
total	$|\mathcal{M}^\star_{\text{FBM}}| \cdot (|\mathcal{M}^\star_L| + 2|\mathcal{M}^\star_{\text{FBM}}| - 1)$	total	$|\mathcal{M}^\star_{\text{FBM}}| \cdot (|P_A| + |\mathcal{M}^\star_{\text{FBM}}|)$

7.3 DEADLOCK PREVENTION POLICY

This section presents a deadlock prevention policy to obtain an optimal liveness-enforcing supervisor with a minimal number of control places if such a supervisor exists.

Algorithm 7.1 Structurally Minimal Deadlock Prevention Policy

Input: Petri net model (N, M_0) of an AMS with $N = (P^0 \cup P_A \cup P_R, T, F, W)$.

Output: An optimally controlled Petri net system (N_1, M_1) with a minimal structure if such a supervisor exists.

1) Compute the set of FBMs $\mathcal{M}_{\text{FBM}}$ and the set of legal markings $\mathcal{M}_L$ for (N, M_0).

2) Compute the minimal covered set of FBMs $\mathcal{M}^\star_{FBM}$ and the minimal covering set of legal markings $\mathcal{M}^\star_L$ for (N, M_0).

3) $V_M := \emptyset$. /* V_M is used to denote the set of control places to be computed.*/

4) Solve the ILPP, i.e., MCPP, presented in Section 7.2. If it has no solution, exit, as the Petri net model has no optimal liveness-enforcing supervisor expressed by control places.

5) **foreach** $\{h_j = 1\}$ **do**

Use $l_{j,i}$'s in the solution as the coefficients of a PI and design a control place p_{c_j} to forbid $M_j \in \mathcal{M}^\star_{FBM}$ by the method presented in Section 5.2.1.

$V_M := V_M \cup \{p_{c_j}\}$.

6) Add all control places in V_M to (N, M_0) and denote the resulting net system as (N_1, M_1).

7) Output (N_1, M_1).

8) End.

This algorithm can obtain an optimal liveness-enforcing supervisor with the minimal number of control places. We first compute the reachability graph of a Petri net model and then the sets of legal markings and FBMs. Using the vector covering approach, the two sets are reduced to be small, i.e., the minimal covering set of legal markings $\mathcal{M}^\star_L$ and the minimal covered set of FBMs $\mathcal{M}^\star_{FBM}$. Then, we design MCPP to select the control places to be added. For each FBM in $\mathcal{M}^\star_{FBM}$, there exists a constraint to ensure that it can be forbidden by at least one PI. Thus, all FBMs are forbidden and the controlled system cannot enter the DZ. The objective function of MCPP is used to minimize the number of control places to be selected. Finally, a minimal number of control places are selected to obtain an optimal liveness-enforcing supervisor. It is not an iterative synthesis policy since it can find all control places by solving an ILPP. The most advantage of this method is that an optimal liveness-enforcing supervisor with the minimal number of control places can be definitely found if such a supervisor exists.

Theorem 7.1 *Algorithm 7.1 can obtain a maximally permissive supervisor with the minimal number of control places for a Petri net model of an AMS if there exists a solution that satisfies Eq. (5.11) for each FBM in* $\mathcal{M}^\star_{FBM}$.

Proof By Eq. (7.9), every FBM in $\mathcal{M}^\star_{FBM}$ is forbidden by at least one control place in the final supervisor. According to Corollary 5.3, if all markings in $\mathcal{M}^\star_{FBM}$ are forbidden, then all FBMs are forbidden, implying that the controlled system is live since it cannot enter the DZ. By Corollary 5.4, if the set of integer linear inequalities of Eq. (5.11) for each FBM has a solution, then no legal markings are forbidden. That is to say, all legal markings are reachable and an optimal supervisor is obtained by Algorithm 7.1. From the objective function of the MCPP, it is known that the minimal number of control places is ensured. In a word, the final supervisor is maximally permissive and has the minimum control places. □

Theorem 7.2 *The presented deadlock prevention method can lead to a maximally permissive liveness-enforcing supervisor with the minimal number of control places if such a supervisor exists.*

Proof The proof of Theorem 5.4 in Chapter 5 shows that for every additional control place, there exists a PI that is associated with it and operation places only. Assume that an optimal supervisor exists for a net model, then each control place of the supervisor can construct a PI that does not forbid any legal marking and forbids at least one FBM. By contradiction, suppose that there exists a PI that does not satisfy Eq. (5.11) for some markings in $M_L^\star$, as Section 5.3 describes, these markings are forbidden by the PI. This means that the control place designed for this PI is not optimal. Thus, it is proved that any PI for the optimal control purposes satisfies Eq. (5.11). According to Theorem 7.1, the presented method can obtain an optimal supervisor with the minimal number of control places if there exists a solution that satisfies Eq. (5.11) for each FBM in $M_{FBM}^\star$. Therefore, the presented method can obtain an optimal liveness-enforcing supervisor with the minimal number of control places if such a supervisor exists.   □

Now we consider the Petri net model of an AMS shown in Fig. 7.1 to show the application of the deadlock prevention approach. In this Petri net model, there are 11 places and eight transitions. The places can be classified into three groups: idle place set $P^0 = \{p_1, p_8\}$, operation place set $P_A = \{p_2 - p_7\}$, and resource place set $P_R = \{p_9, p_{10}, p_{11}\}$. The model has 20 reachable markings, 15 of which are legal and five are FBMs. Using the vector covering approach, the minimal covered set of FBMs and the minimal covering set of legal markings have three and two elements, respectively, i.e., $M_{FBM}^\star = \{p_2 + p_5, p_3 + p_5, p_2 + p_6\}$ and $M_L^\star = \{p_2 + p_3 + p_4, p_5 + p_6 + p_7\}$.

For $FBM_1 = p_2 + p_5$, let I_1 be the PI to forbid FBM_1, whose coefficients are $l_{1,i}$'s ($i \in \{2,3,4,5,6,7\}$). For the optimal control purposes, I_1 must not forbid any marking in $M_L^\star$. Thus, I_1 should satisfy the two constraints: $l_{1,2} \cdot (1-1) + l_{1,3} \cdot (1-0) + l_{1,4} \cdot (1-0) + l_{1,5} \cdot (0-1) + l_{1,6} \cdot (0-0) + l_{1,7} \cdot (0-0) \leq -1$ and $l_{1,2} \cdot (0-1) + l_{1,3} \cdot (0-0) + l_{1,4} \cdot (0-0) + l_{1,5} \cdot (1-1) + l_{1,6} \cdot (1-0) + l_{1,7} \cdot (1-0) \leq -1$ for the two legal markings $p_2 + p_3 + p_4$ and $p_5 + p_6 + p_7$ in $M_L^\star$, respectively. By simplifying the two constraints, we have:

$$l_{1,3} + l_{1,4} - l_{1,5} \leq -1 \quad \text{and}$$

$$-l_{1,2} + l_{1,6} + l_{1,7} \leq -1.$$

We introduce two variables $f_{1,2}$ and $f_{1,3}$ ($f_{1,2}, f_{1,3} \in \{0,1\}$) to represent whether I_1 forbids $FBM_2 = p_3 + p_5$ and $FBM_3 = p_2 + p_6$, respectively. Thus, we have the two constraints, i.e., $l_{1,2} \cdot (0-1) + l_{1,3} \cdot (1-0) + l_{1,4} \cdot (0-0) + l_{1,5} \cdot (1-1) + l_{1,6} \cdot (0-0) + l_{1,7} \cdot (0-0) \geq -Q \cdot (1-f_{1,2})$ and $l_{1,2} \cdot (1-1) + l_{1,3} \cdot (0-0) + l_{1,4} \cdot (0-0) + l_{1,5} \cdot (0-1) + l_{1,6} \cdot (1-0) + l_{1,7} \cdot (0-0) \geq -Q \cdot (1-f_{1,3})$, where Q is a positive constant integer that must be big enough. For the former constraint, $f_{1,2} = 1$ indicates that FBM_2 is forbidden by I_1 and $f_{1,2} = 0$ indicates that this constraint is redundant and FBM_2 cannot be forbidden by I_1. Similarly, for the latter constraint, $f_{1,3} = 1$ indicates that FBM_3 is forbidden by I_1 and $f_{1,3} = 0$ indicates that this constraint is redundant and FBM_3 cannot be forbidden by I_1. By simplifying the two constraints, we have:

$$-l_{1,2} + l_{1,3} \geq -Q \cdot (1-f_{1,2}) \quad \text{and}$$

$$-l_{1,5} + l_{1,6} \geq -Q \cdot (1-f_{1,3}).$$

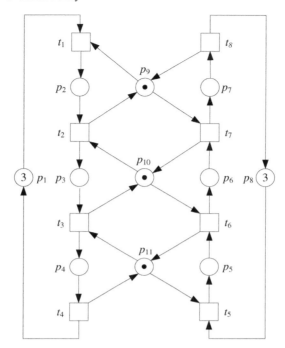

Figure 7.1 Petri net model of an AMS.

Similarly, for FBM$_2$, we have the following constraints:

$$l_{2,2} + l_{2,4} - l_{2,5} \leq -1 \quad \text{and}$$
$$-l_{2,3} + l_{2,6} + l_{2,7} \leq -1,$$

and
$$l_{2,2} - l_{2,3} \geq -Q \cdot (1 - f_{2,1}) \quad \text{and}$$
$$l_{2,2} - l_{2,3} - l_{2,5} + l_{2,6} \geq -Q \cdot (1 - f_{2,3}).$$

Finally, for FBM$_3$, we have:

$$l_{3,3} + l_{3,4} - l_{3,6} \leq -1 \quad \text{and}$$
$$-l_{3,2} + l_{3,5} + l_{3,7} \leq -1,$$

and
$$l_{3,5} - l_{3,6} \geq -Q \cdot (1 - f_{3,1}) \quad \text{and}$$
$$-l_{3,2} + l_{3,3} + l_{3,5} - l_{3,6} \geq -Q \cdot (1 - f_{3,2}).$$

Now we introduce a set of variables h_j's ($j \in \{1, 2, 3\}$), where $h_j{=}1$ indicates that I_j is selected to compute a control place and $h_j{=}0$ indicates that I_j is not selected. FBM$_1$ must be forbidden by at least one PI. Thus, we have the following constraint:

$$h_1 + f_{2,1} + f_{3,1} \geq 1.$$

Similarly, for FBM_2 and FBM_3, we have:

$$h_2 + f_{1,2} + f_{3,2} \geq 1$$

and

$$h_3 + f_{1,3} + f_{2,3} \geq 1.$$

If I_1 is not selected to compute a control place, i.e., $h_1 = 0$, it cannot forbid FBM_2 and FBM_3. Therefore, we obtain:

$$f_{1,2} \leq h_1 \text{ and}$$
$$f_{1,3} \leq h_1.$$

Similarly, for I_2 and I_3, we derive:

$$f_{2,1} \leq h_2 \text{ and}$$
$$f_{2,3} \leq h_2,$$

and

$$f_{3,1} \leq h_3 \text{ and}$$
$$f_{3,2} \leq h_3.$$

Given all the above constraints, now we use an objective function to minimize the number of control places to be computed, as shown below:

$$\min \ h_1 + h_2 + h_3.$$

Grouping all the constraints above and the objective function, an ILPP, i.e., MCPP, is formulated as follows:

MCPP:

$$\min \ h_1 + h_2 + h_3$$

subject to

$$l_{1,3} + l_{1,4} - l_{1,5} \leq -1$$
$$-l_{1,2} + l_{1,6} + l_{1,7} \leq -1$$
$$-l_{1,2} + l_{1,3} \geq -Q \cdot (1 - f_{1,2})$$
$$-l_{1,5} + l_{1,6} \geq -Q \cdot (1 - f_{1,3})$$
$$l_{2,2} + l_{2,4} - l_{2,5} \leq -1$$
$$-l_{2,3} + l_{2,6} + l_{2,7} \leq -1$$
$$l_{2,2} - l_{2,3} \geq -Q \cdot (1 - f_{2,1})$$

$$l_{2,2} - l_{2,3} - l_{2,5} + l_{2,6} \geq -Q \cdot (1 - f_{2,3})$$

$$l_{3,3} + l_{3,4} - l_{3,6} \leq -1$$

$$-l_{3,2} + l_{3,5} + l_{3,7} \leq -1$$

$$l_{3,5} - l_{3,6} \geq -Q \cdot (1 - f_{3,1})$$

$$-l_{3,2} + l_{3,3} + l_{3,5} - l_{3,6} \geq -Q \cdot (1 - f_{3,2})$$

$$h_1 + f_{2,1} + f_{3,1} \geq 1$$

$$h_2 + f_{1,2} + f_{3,2} \geq 1$$

$$h_3 + f_{1,3} + f_{2,3} \geq 1$$

$$f_{1,2} \leq h_1$$

$$f_{1,3} \leq h_1$$

$$f_{2,1} \leq h_2$$

$$f_{2,3} \leq h_2$$

$$f_{3,1} \leq h_3$$

$$f_{3,2} \leq h_3$$

$$l_{j,i} \in \{0, 1, 2, \cdots\}, \ \forall i \in \{2, 3, 4, 5, 6, 7\} \text{ and } \forall j \in \{1, 2, 3\}$$

$$f_{j,k} \in \{0, 1\}, \ \forall j, k \in \{1, 2, 3\} \text{ and } j \neq k$$

$$h_j \in \{0, 1\}, \ \forall j \in \{1, 2, 3\}$$

The above MCPP has a solution as shown in Table 7.2. In the solution, since $h_1 = 0$, $h_2 = 1$, and $h_3 = 1$, I_2 and I_3 are selected to compute the control places and there is no need to compute a control place for I_1 to forbid FBM$_1$. In fact, $f_{2,1} = 1$ in the solution means that FBM$_1$ is forbidden by I_2. There are totally two control places to be added for this net, as shown in Table 7.3, where the first column is the selected FBM number i, the second represents the selected first-met bad marking FBM$_i$, the third shows the PI I_i, and the fourth to the sixth columns indicate pre-transitions ($^\bullet p_{c_i}$), post-transitions ($p_{c_i}^\bullet$), and initial marking ($M_0(p_{c_i})$) of control place p_{c_i}, respectively. Adding the two control places to the original net model, as shown in Fig. 7.2, the resulting net is live with 15 legal markings. Note that in this solution, though I_1 that is designed to forbid FBM$_1$ is not selected, FBM$_1$ is forbidden by I_2 since $f_{2,1} = 1$. In fact, we can substitute FBM$_1$ into I_2: $\mu_2 + \mu_3 + 2\mu_5 \leq 2$. Then, we have FBM$_1(p_2)$+FBM$_1(p_3)$+2FBM$_1(p_5) = 3 \nleq 2$, which contradicts the constraint of I_2. Thus, FBM$_1$ is forbidden by I_2 and there is no need to compute a control place for I_1.

Table 7.2 A solution of the MCPP for the net in Fig. 7.1

Variable	h_1	$f_{1,2}$	$f_{1,3}$	$l_{1,2}$	$l_{1,3}$	$l_{1,4}$	$l_{1,5}$	$l_{1,6}$	$l_{1,7}$
Value	0	0	0	1	0	0	1	0	0
Variable	h_2	$f_{2,1}$	$f_{2,3}$	$l_{2,2}$	$l_{2,3}$	$l_{2,4}$	$l_{2,5}$	$l_{2,6}$	$l_{2,7}$
Value	1	1	0	1	1	0	2	0	0
Variable	h_3	$f_{3,1}$	$f_{3,2}$	$l_{3,2}$	$l_{3,3}$	$l_{3,4}$	$l_{3,5}$	$l_{3,6}$	$l_{3,7}$
Value	1	0	0	1	0	0	0	1	0

Table 7.3 Control places computed for the net shown in Fig. 7.1

i	FBM_i	l_i	$\bullet p_{c_i}$	$p_{c_i}^{\bullet}$	$M_0(p_{c_i})$
2	$p_3 + p_5$	$\mu_2 + \mu_3 + 2\mu_5 \leq 2$	$t_3, 2t_6$	$t_1, 2t_5$	2
3	$p_2 + p_6$	$\mu_2 + \mu_6 \leq 1$	t_2, t_7	t_1, t_6	1

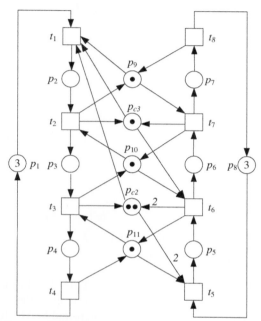

Figure 7.2 An optimally controlled system of the Petri net in Fig. 7.1.

7.4 EXPERIMENTAL RESULTS

In this section, AMS examples available in the literature are considered to show the experimental results of the presented approach.

The Petri net model of an AMS from (Huang *et al.*, 2006) is shown in Fig. 7.3. There are 16 places and 13 transitions. The places have the following set partitions: $P^0 = \{p_1, p_8\}$, $P_R = \{p_{13}, \ldots, p_{16}\}$, and $P_A = \{p_2, \ldots, p_7, p_9, \ldots, p_{12}\}$. It has 77 reachable markings, 13 and 64 of which are FBMs and legal ones, respectively. By using the vector covering approach, $\mathcal{M}_{FBM}^{\star}$ and $\mathcal{M}_L^{\star}$ have five and 12 markings, respectively. For this example, the MCPP has 105 constraints and 75 variables. By solving the MCPP, we obtain an optimal solution in one second with two control places only that need to be added, as shown in Table 7.4.

Table 7.5 shows a few available results for the example in terms of the numbers of additional places, additional arcs, and reachable markings of the controlled net. It can be seen that the results listed in the table are optimal but the presented method

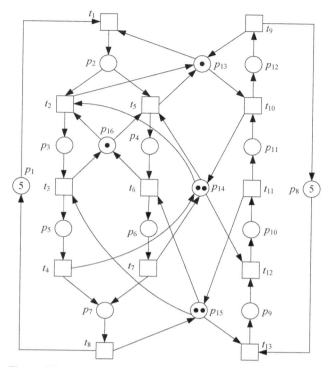

Figure 7.3 Petri net model of an AMS from (Huang *et al.*, 2006).

Table 7.4 Control places computed for the net shown in Fig. 7.3

i	FBM_i	I_i	$^\bullet p_{c_i}$	$p_{c_i}^\bullet$	$M_0(p_{c_i})$
2	$p_2 + p_{10}$	$\mu_2 + \mu_{10} + \mu_{11} \leq 1$	t_2, t_5, t_{10}	t_1, t_{12}	1
5	$p_3 + 2p_9$	$\mu_2 + \mu_3 + \mu_4 + 2\mu_9 \leq 4$	$t_3, t_6, 2t_{12}$	$t_1, 2t_{13}$	4

obtains a supervisor with two control places and 10 arcs only, both of which are the smallest compared with the related numbers generated by other existing methods.

Table 7.5 Performance comparison of typical deadlock control policies

Parameters	(Huang *et al.*, 2006)	(Uzam and Zhou, 2006)	Alg. 7.1
No. of monitors	5	5	2
No. of arcs	27	23	10
No. of states	64	64	64

The Petri net model of an AMS from (Uzam, 2002) is shown in Fig. 7.4. There are 19 places and 14 transitions. The places have the following set partitions: $P^0 = \{p_1, p_8\}$, $P_R = \{p_{14}, \ldots, p_{19}\}$, and $P_A = \{p_2, \ldots, p_7, p_9, \ldots, p_{13}\}$. It has 282 reachable markings, 54 and 205 of which are FBMs and legal ones, respectively. By using the vector covering approach, $\mathcal{M}_{FBM}^\star$ and $\mathcal{M}_L^\star$ have eight and 26 markings, respectively. For this example, the MCPP has 328 constraints and 152 variables. By solving the MCPP, we obtain a solution such that only two control places need to be added, as shown in Table 7.6.

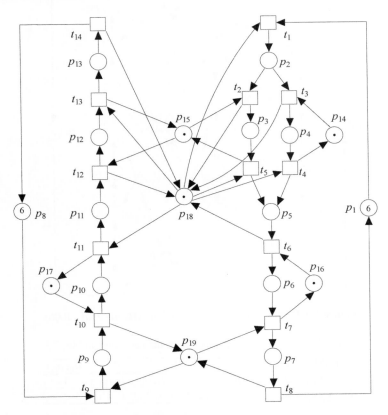

Figure 7.4 Petri net model of an AMS from (Uzam, 2002).

Table 7.6 Control places computed for the net shown in Fig. 7.4

i	FBM_i	I_i	$^\bullet p_{c_i}$	$p_{c_i}^\bullet$	$M_0(p_{c_i})$
5	$p_2 + p_4 + p_6 +$ $p_9 + p_{10}$	$\mu_2 + 2\mu_3 + \mu_4 + 2\mu_5 +$ $2\mu_6 + 3\mu_9 + 3\mu_{10} \leq 9$	$2t_7, 3t_{11}$	$t_1, t_2, t_4, 3t_9$	9
7	$p_2 + p_4 + p_{12}$	$\mu_2 + 2\mu_3 + \mu_4 + 2\mu_{11} +$ $2\mu_{12} \leq 3$	$t_4, 2t_5, 2t_{13}$	$t_1, t_2, 2t_{11}$	3

This example is widely used in the literature. Table 7.7 shows some available results for the example in terms of the numbers of additional control places and arcs, and the reachable markings of the controlled net. It can be seen that the results listed in the table are optimal but the presented method obtains a supervisor with only two control places and 12 arcs, both of which are minimal compared with the methods in the literature.

Table 7.7 Performance comparison of typical deadlock control policies

Parameters	(Uzam, 2002)	(Li et al., 2008)	(Piroddi et al., 2008)	Alg. 5.3	Alg. 7.1
No. monitors	6	9	5	8	2
No. arcs	32	42	23	37	12
No. states	205	205	205	205	205

The Petri net model of an AMS from (Ezpeleta et al., 1995) is shown in Fig. 7.5. There are 26 places and 20 transitions. Its places have the following set partitions: $P^0 = \{p_1, p_5, p_{14}\}$, $P_R = \{p_{20}, \ldots, p_{26}\}$, and $P_A = \{p_2, \ldots, p_4, p_6, \ldots, p_{13}, p_{15}, \ldots, p_{19}\}$. It has 26,750 reachable markings, 4,211 and 21,581 of which are FBMs and legal ones, respectively. By using the vector covering approach, $\mathcal{M}^\star_{FBM}$ and $\mathcal{M}^\star_L$ have 34 and 393 markings, respectively. For this example, the MCPP has 15,640 constraints and 1,700 variables. By solving the MCPP, we obtain a solution with five control places, as shown in Table 7.8.

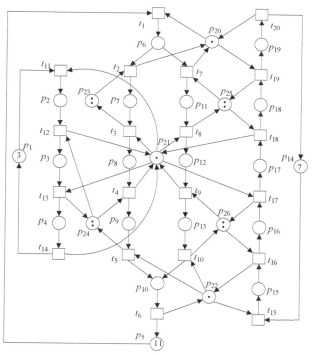

Figure 7.5 An S^3PR model in (Ezpeleta et al., 1995).

This example is also studied in a number of papers. Table 7.9 shows some classical available results for the example in terms of the numbers of the additional places and the arcs, and the reachable markings of the controlled net. It can be seen that, among all the six approaches shown in this table, the method in this chapter obtains a supervisor with five control places and 55 arcs only. Compared with other optimal deadlock prevention policies, the numbers of additional control places and arcs derived from the method are minimum.

Table 7.8 Control places computed for the net shown in Fig. 7.5

i	FBM$_i$	I_i	$\bullet p_{c_i}$	$p_{c_i}\bullet$	$M_0(p_{c_i})$
4	$p_{12}+2p_{16}$	$8\mu_2 + 8\mu_3 + 3\mu_6 + \mu_7 + 9\mu_8 + 9\mu_9 + 2\mu_{11} + 35\mu_{12} + 9\mu_{15} + 35\mu_{16} + \mu_{17} \leq 104$	$2t_2, 9t_5, t_7, 35t_9, 8t_{13}, 34t_{17}, t_{18}$	$3t_1, 8t_3, 33t_8, 8t_{11}, 9t_{15}, 26t_{16}$	104
15	$p_6 + 2p_7 + 2p_9 + p_{11} + p_{12} + p_{15} + p_{16} + p_{18}$	$17\mu_6 + 17\mu_7 + 2\mu_8 + 2\mu_9 + 17\mu_{11} + 3\mu_{12} + 3\mu_{13} + 4\mu_{15} + 4\mu_{16} + 15\mu_{17} + 17\mu_{18} \leq 99$	$2t_5, 15t_3, 14t_8, 3t_{10}, 17t_{19}$	$17t_1, 4t_{15}, 11t_{17}, 2t_{18}$	99
18	$p_3 + p_8 + p_9 + 2p_{11} + p_{15} + p_{16}$	$\mu_2 + \mu_3 + \mu_8 + \mu_9 + 39\mu_{11} + 3\mu_{12} + 3\mu_{13} + 6\mu_{15} + 15\mu_{16} + 24\mu_{17} \leq 101$	$t_5, 36t_8, 3t_{10}, t_{13}, 24t_{18}$	$t_3, 39t_7, t_{11}, 6t_{15}, 9t_{16}, 9t_{17}$	101
25	$p_2 + 2p_3$	$28\mu_2 + 28\mu_3 + 3\mu_7 + 28\mu_8 + 7\mu_9 + 7\mu_{13} + 7\mu_{15} + 7\mu_{16} \leq 83$	$21t_4, 7t_5, 7t_{10}, 28t_{13}, 7t_{17}$	$3t_2, 25t_3, 7t_9, 28t_{11}, 7t_{15}$	83
33	$2p_{13} + p_{15}$	$5\mu_2 + \mu_6 + \mu_7 + 5\mu_8 + 5\mu_9 + \mu_{11} + 15\mu_{12} + 31\mu_{13} + 41\mu_{15} + 16\mu_{16} + \mu_{17} \leq 102$	$5t_5, 31t_{10}, 5t_{12}, 25t_{16}, 15t_{17}, t_{18}$	$t_1, 4t_3, 14t_8, 16t_9, 5t_{11}, 41t_{15}$	102

Table 7.9 Performance comparison of typical deadlock control policies

Parameters	(Ezpeleta et al., 1995)	(Li and Zhou, 2004)	(Huang et al., 2001)	(Uzam and Zhou, 2006)	(Piroddi et al., 2008)	Alg. 5.3	Alg. 7.1
No. monitors	18	6	16	19	13	17	5
No. arcs	106	32	88	112	82	101	55
No. states	6287	6287	12656	21562	21581	21581	21581

7.5 CONCLUSIONS

This chapter presents a deadlock prevention policy to obtain an optimal supervisor with the minimal number of control places if such a supervisor exists. The performance analysis of a deadlock prevention policy is always carried out by considering behavioral permissiveness, structural complexity, and computational complexity. For the presented method, in terms of behavioral permissiveness, the derived supervisor is optimal. In terms of structural complexity, it has the smallest number of control places among all the optimal ones in the literature and more importantly, no redundant control place can survive in the supervisor. In fact, for the examples in Figs. 7.4 and 7.5, the numbers of the additional control places are two and five, respectively. However, the numbers of elementary siphons (which is linearly dependent on the size of the net) for the two examples are three and six, respectively. This means that the presented method is possible to obtain a less number of control places than elementary siphon approaches.

The only problem of the presented method is its computational complexity since one requires a complete enumeration of reachable markings and needs to solve an ILPP. The number of reachable markings increases exponentially with respect to the size of a net, which always suffers from the state explosion problem. The number of constraints in the ILPP is polynomial with respect to the number of elements in the minimal covering set of legal markings and the minimal covered set of FBMs. To reduce the computational costs, BDDs can be used (Pastor *et al.*, 1994: Pastor, *et al.*, 2001) to compute the reachable markings. For instance, in Chapters 3 and 5, a number of algorithms are developed to compute the set of legal markings, the set of FBMs, the minimal covering set of legal markings, and the minimal covered set of FBMs by using BDDs. Considering that an ILPP is NP-hard, a vector covering approach is used to greatly decrease the number of constraints in the ILPP. However, the method is still inapplicable to large-scale Petri net models if there are a lot of constraints.

7.6 BIBLIOGRAPHICAL REMARKS

Many researchers try to design liveness-enforcing Petri net supervisors with as many legal markings and less control places as possible. Li and Zhou (2004) develop the concept of elementary siphons and formulate a systematic method such that a liveness-enforcing supervisor can be obtained by adding control places for elementary siphons only, implying that the number of control places is linearly dependent on the size of a net (Reveliotis, 2007). Although structural complexity is remarkably reduced, their elementary siphon approach cannot in general produce a maximally permissive supervisor. Piroddi *et al.* (2008, 2009) propose a method by using a set covering approach, which can lead to an almost maximally permissive (in fact it is maximally permissive for every example in (Piroddi *et al.*, 2008, 2009) but they cannot prove it) supervisor with a very simple supervisory structure.

For a well-known Petri net model depicted in Fig. 7.5, they first find its optimal supervisor with 13 control places. In this chapter, we provide an optimal supervisor for this example with five control places only. Most materials in this chapter refer to (Chen and Li, 2011).

References

Chen, Y. F. and Z. W. Li. 2011. Design of a maximally permissive liveness-enforcing supervisor with a compressed supervisory structure for flexible manufacturing systems. Automatica. 47(5): 1028–1034.

Ezpeleta, J., J. M. Colom, and J. Martinez. 1995. A Petri net based deadlock prevention policy for flexible manufacturing systems. IEEE Transactions on Robotics and Automation. 11(2): 173–184.

Ezpeleta, J., F. Tricas, F. Garcia-Valles, and J. M. Colom. 2002. A banker's solution for deadlock avoidance in FMS with flexible routing and multiresource states. IEEE Transactions on Robotics and Automation. 18(4): 621–625.

Fanti, M. P. and M. C. Zhou. 2004. Deadlock control methods in automated manufacturing systems. IEEE Transactions on Systems, Man, and Cybernetics, Part A. 34(1): 5–22.

Fanti, M. P. and M. C. Zhou. Deadlock control methods in automated manufacturing systems. pp. 1–22. In M. C. Zhou and M. P. Fanti. [eds.]. 2005. Deadlock Resolution in Computer-Integrated Systems. New York: Marcel-Dekker Co.

Ghaffari, A., N. Rezg, and X. L. Xie. 2003. Design of a live and maximally permissive Petri net controller using the theory of regions. IEEE Transactions on Robotics and Automation. 19(1): 137–142.

Hsieh, F. S. and S. C. Chang. 1994. Dispatching-driven deadlock avoidance controller synthesis for flexible manufacturing systems. IEEE Transactions on Robotics and Automation. 10(2): 196–209.

Huang, Y. S., M. D. Jeng, X. L. Xie, and S. L. Chung. 2001. Deadlock prevention based on Petri nets and siphons. International Journal of Production Research. 39(2): 283–305.

Huang, Y. S., M. D. Jeng, X. L.Xie, and D. H. Chung. 2006. Siphon-based deadlock prevention for flexible manufacturing systems. IEEE Transactions on Systems, Man, and Cybernetics, Part A. 36(6): 1248–1256.

Jeng, M. D. and X. L. Xie. Deadlock detection and prevention of automated manufacturing systems using Petri nets and siphons. pp. 233–281. In M. C. Zhou and M. P. Fanti. [eds.]. 2005. Deadlock Resolution in Computer-Integrated Systems. New York: Marcel-Dekker Inc. Co.

Lautenbach, K. and H. Ridder. 1994. Liveness in bounded Petri nets which are covered by T-invariants. Lecture Notes in Computer Science. 815: 358–375.

Li, Z. W. and M. C. Zhou. 2004. Elementary siphons of Petri nets and their application to deadlock prevention in flexible manufacturing systems. IEEE Transactions on Systems, Man, and Cybernetics, Part A. 34(1): 38–51.

Li, Z. W. and M. C. Zhou. 2009. Deadlock Resolution in Automated Manufacturing Systems: A Novel Petri Net Approach. Berlin: Springer-Verlag.

Li, Z. W., M. C. Zhou, and M. D. Jeng. 2008. A maximally permissive deadlock prevention policy for FMS based on Petri net siphon control and the theory of regions. IEEE Transactions on Automation Science and Engineering. 5(1): 182–188.

Murata, T. 1989. Petri nets: Properties, analysis and application. Proceedings of the IEEE. 77(4): 541–580.

Park, J. and S. A. Reveliotis. 2000. Algebraic synthesis of efficient deadlock avoidance policies for sequential resource allocation systems. IEEE Transactions on Robotics and Automation. 16(2): 190–195.

Park, J. and S. A. Reveliotis. 2001. Deadlock avoidance in sequential resource allocation systems with multiple resource acquisitions and flexible routings. IEEE Transactions on Automatic Control. 46(10): 1572–1583.

Pastor, E., O. Roig, J. Cortadella, and R. M. Badia. 1994. Petri net analysis using Boolean manipulation. Lecture Notes in Computer Science. 815: 416–435.

Pastor, E., J. Cortadella, and O. Roig. 2001. Symbolic analysis of bounded Petri nets. IEEE Transactions on Computers. 50(5): 432–448.

Piroddi, L., R. Cordone, and I. Fumagalli. 2008. Selective siphon control for deadlock prevention in Petri nets. IEEE Transactions on Systems, Man, and Cybernetics, Part A. 38(6): 1337–1348.

Piroddi, L., R. Cordone, and I. Fumagalli. 2009. Combined siphon and marking generation for deadlock prevention in Petri nets. IEEE Transactions on Systems, Man, and Cybernetics, Part A. 39(3): 650–661.

Reveliotis, S. A. 2007. Implicit siphon control and its role in the liveness-enforcing supervision of sequential resource allocation systems. IEEE Transactions on Systems, Man, and Cybernetics, Part A. 37(3): 319–328.

Tricas, F., F. Garcia-Valles, J. M. Colom, and J. Ezpelata. 1998. A structural approach to the problem of deadlock prevention in processes with resources, 273–278. In Proceedings of WODES'98: Italy, 26–28 August.

Tricas, F., F. Garcia-Valles, J. M. Colom, and J. Ezpelata. An iterative method for deadlock prevention in FMS. pp. 139–148. In G. Stremersch [ed.]. 2000. Discrete Event Systems: Analysis and Control. Kluwer Academic: Boston, MA.

Uzam, M. 2002. An optimal deadlock prevention policy for flexible manufacturing systems using Petri net models with resources and the theory of regions. International Journal of Advanced Manufacturing Technology. 19(3): 192–208.

Uzam, M. and M. C. Zhou. 2006. An improved iterative synthesis method for liveness enforcing supervisors of flexible manufacturing systems. International Journal of Production Research. 44(10): 1987–2030.

Xing, K. Y., B. S. Hu, and H. X. Chen. Deadlock avoidance policy for flexible manufacturing systems. pp. 239–263. In M. C. Zhou. [ed.]. 1995. Petri Nets in Flexible and Agile Automation. Kluwer Academic: Boston, MA.

Yamalidou, K., J. Moody, M. Lemmon, and P. Antsaklis. 1996. Feedback control of Petri nets based on place invariants. Automatica. 32(1): 15–28.

Chapter 8

Behaviorally Optimal and Structurally Simple Supervisors

ABSTRACT

This chapter presents two iterative deadlock prevention policies for AMSs, aiming to design maximally permissive liveness-enforcing supervisors with a small number of control places and high computational efficiency. The vector covering approach is used to obtain a minimal covering set of legal markings and a minimal covered set of FBMs, which are much smaller than the sets of legal markings and FBMs, respectively. At each iteration, by solving an ILPP, a PI associated with a control place is designed to forbid as many FBMs as possible and no markings in the minimal covering set of legal markings are forbidden. The objective function of the ILPP maximizes the number of FBMs that are forbidden by the PI. Then, the forbidden FBMs by the PI are removed from the minimal covered set of FBMs. This process is carried out until all FBMs are forbidden. Two ILPP design techniques are introduced for the control policies. Although the presented methods cannot in general guarantee the minimality of the supervisory structure, they reduce the overall computational time greatly, which is shown by numerical studies. Finally, a number of AMS examples from the literature are utilized to illustrate the methods.

8.1 INTRODUCTION

Last chapter introduces an approach that can obtain a maximally permissive liveness-enforcing supervisor with the minimal number of control places. It is a non-iterative approach since all control places can be obtained by solving an ILPP once (which is denoted as MCPP in Chapter 7). Though this approach overcomes the problems of both behavioral permissiveness and structural complexity, it still suffers from expensive computational costs. Solving an ILPP is NP-hard in theory and greatly depends on the number of constraints and variables in it. For a large-scale Petri net model, the ILPP designed by this approach may have a huge number of constraints and integer variables, which cannot be solved within a reasonable time. For instance, in Chapter 7, the ILPP for a Petri net model from (Ezpeleta *et al.*, 1995) with 26 places and 20 transitions has 15,640 constraints and 1,700 variables. It takes 44 hours to solve it.

This chapter provides an iterative deadlock prevention policy and its a modified version, aiming to overcome the computational complexity problem in Chapter 7 and ensure that the controlled system is maximally permissive with a simple control structure. First, the vector covering approach is used to compute the minimal covering set of legal markings and the minimal covered set of FBMs. Then, an iterative control place design process is carried out. At each iteration, a PI is designed to forbid as many FBMs as possible and the coefficients of the PI are computed by solving an ILPP that satisfies the following two conditions:

1. No markings in the minimal covering set of legal markings are prohibited. Therefore, no legal markings are forbidden, implying that the final supervisor is maximally permissive.
2. The objective function maximizes the number of FBMs that are forbidden by a PI. That is to say, the PI forbids as many FBMs in the minimal covered set of FBMs as possible.

Solving the ILPP can obtain a PI and a control place. All FBMs that are forbidden by the PI are removed from the minimal covered set of FBMs. This process is carried out until no FBM is left in the minimal covered set. A maximally permissive supervisor with a small number of control places can be accordingly obtained.

We also provide a modified version of the above mentioned method in order to further reduce the computational time. In fact, we only modify the iteration steps as follows. At each iteration, an FBM is singled out from the minimal covered set of FBMs and a PI is designed to forbid the selected FBM. The PI is computed by solving an ILPP such that the selected FBM is forbidden and no legal markings in the minimal covering set of legal markings are forbidden. At the same time, we also use an objective function to maximize the number of FBMs that are forbidden by the PI. The modified version of an ILPP has a small number of constraints and variables. Therefore, compared with the former, it can be solved in a shorter time in general.

The two methods are motivated by the work in (Chen *et al.*, 2011) and (Chen and Li, 2011), which have the same application scope as the approaches in (Chen *et al.*, 2011) and (Chen and Li, 2011). That is to say, they can be applied to all AMS-oriented classes of Petri net models in the literature, such as PPN (Banaszak and Krogh, 1990; Hsieh and Chang, 1994; Xing *et al.*, 1995), S^3PR (Ezpeleta *et al.*, 1995), ES^3PR (Tricas *et al.*, 1998), S^4PR (Tricas *et al.*, 2000), S^*PR (Ezpeleta *et al.*, 2002), S^2LSPR (Park and Reveliotis, 2000), S^3PGR^2 (Park and Reveliotis, 2001), and S^3PMR (Huang *et al.*, 2006).

8.2 CONTROL PLACE SYNTHESIS FOR FBMs

This section presents two techniques to design a PI that can forbid as many FBMs as possible. In what follows, $\mathbb{N}^\star_{FBM}$ is used to denote $\{i | M_i \in \mathcal{M}^\star_{FBM}\}$.

Using the method presented in Chapter 5, we can design a maximally permissive PI to forbid a given FBM. In fact, a PI may forbid more FBMs. Now we report a technique to maximize the number of FBMs forbidden by a PI. Let I be a PI for the constraint

$$\sum_{i \in \mathbb{N}_A} l_i \cdot \mu_i \leq \beta \qquad (8.1)$$

where l_i's ($i \in \mathbb{N}_A$) are the coefficients of I and β is a positive integer variable.

An FBM $M \in \mathcal{M}_{\text{FBM}}^{\star}$ is forbidden by I if

$$\sum_{i \in \mathbb{N}_A} l_i \cdot M(p_i) \geq \beta + 1 \qquad (8.2)$$

We introduce a set of variables f_k's ($k = 1, 2, \ldots$) to represent the relation between I and M_k in $\mathcal{M}_{\text{FBM}}^{\star}$. Then, Eq. (8.2) is modified as:

$$\sum_{i \in \mathbb{N}_A} l_i \cdot M_k(p_i) \geq \beta + 1 - Q \cdot (1 - f_k) \qquad (8.3)$$

where Q is a positive integer constant that must be big enough and $f_k \in \{0, 1\}$. In Eq. (8.3), $f_k = 1$ indicates that M_k is forbidden by I and $f_k = 0$ indicates that this constraint is redundant and M_k cannot be forbidden by I. All constraints for FBMs in $\mathcal{M}_{\text{FBM}}^{\star}$ are grouped as follows:

$$\sum_{i \in \mathbb{N}_A} l_i \cdot M_k(p_i) \geq \beta + 1 - Q \cdot (1 - f_k), \ \forall M_k \in \mathcal{M}_{\text{FBM}}^{\star} \qquad (8.4)$$

The coefficients of I and β should satisfy the reachability conditions. As a result, we have the following ILPP to design a PI, which is denoted as the Maximal number of Forbidding FBM Problem 1 (MFFP1).

MFFP1:

$$\max \ f = \sum_{k \in \mathbb{N}_{\text{FBM}}^{\star}} f_k$$

subject to

$$\sum_{i \in \mathbb{N}_A} l_i \cdot M_l(p_i) \leq \beta, \ \forall M_l \in \mathcal{M}_L^{\star} \qquad (8.5)$$

$$\sum_{i \in \mathbb{N}_A} l_i \cdot M_k(p_i) \geq \beta + 1 - Q \cdot (1 - f_k), \ \forall M_k \in \mathcal{M}_{\text{FBM}}^{\star} \qquad (8.6)$$

$$l_i \in \{0, 1, 2, \ldots\}, \ \forall i \in \mathbb{N}_A$$

$$\beta \in \{1, 2, \ldots\}$$

$$f_k \in \{0, 1\}, \ \forall k \in \mathbb{N}_{\text{FBM}}^{\star}$$

The objective function f is used to maximize the number of FBMs that are forbidden by I. Denote its optimal value by f^*. If $f^* = 0$, we have $f_k = 0, \ \forall k \in \mathbb{N}_{\text{FBM}}^{\star}$, implying that no FBMs in $\mathcal{M}^{\star}_{\text{FBM}}$ can be forbidden by a maximally permissive PI. In this case, we claim that there is no maximally permissive PI for any FBM in $\mathcal{M}^{\star}_{\text{FBM}}$, as stated below.

Theorem 8.1 *If $f^* = 0$, there is no maximally permissive PI for any FBM in $\mathcal{M}^{\star}_{\text{FBM}}$.*

Proof By contradiction, suppose that there exists a maximally permissive PI I that can forbid an FBM $M_k \in \mathcal{M}^{\star}_{\text{FBM}}$. Since I is maximally permissive, its coefficients $l_1, l_2, \ldots$, and $l_{\mathbb{N}_A}$ satisfy Eq. (8.5). Since M_k is forbidden by I, we have $\sum_{i \in \mathbb{N}_A} l_i \cdot M_k(p_i) \geq \beta + 1$. Therefore, $f_k = 1$ satisfies Eq. (8.6). We have $f^* = \sum_{k \in \mathbb{N}^{\star}_{\text{FBM}}} f_k \geq 1$. This contradicts $f^* = 0$. Thus the conclusion holds. □

As known, solving an ILPP is NP-hard. However, its computational time greatly depends on the number of constraints and variables in it. Therefore, we discuss MFFP1 from the viewpoint of the number of constraints and variables. The numbers of the constraints in Eqs.(8.5) and (8.6) are $|\mathcal{M}^{\star}_L|$ and $|\mathcal{M}^{\star}_{\text{FBM}}|$, respectively. Thus, the total number of the constraints in MFFP1 is $|\mathcal{M}^{\star}_{\text{FBM}}| + |\mathcal{M}^{\star}_L|$. Now we consider the number of variables. Let $|P_A|$ be the number of operation places. The number of variables l_i's $(i \in \mathbb{N}_A)$ is $|P_A|$. Since $k \in \mathbb{N}^{\star}_{\text{FBM}}$, the number of variables f_k's is $|\mathcal{M}^{\star}_{\text{FBM}}|$. Finally, considering the single variable β, MFFP1 has $|P_A| + |\mathcal{M}^{\star}_{\text{FBM}}| + 1$ variables in total.

In fact, a PI to be computed should forbid at least one FBM. Thus, an FBM can be forbidden through setting β as a linear combination of l_i's $(i \in \mathbb{N}_A)$ by Eq. (5.8). In what follows, we modify MFFP1 in order to improve its efficiency by slightly reducing the number of its variables and constraints. Let I be a PI that forbids an FBM $M_j \in \mathcal{M}^{\star}_{\text{FBM}}$ by Eq. (5.7). $\forall M_k \in \mathcal{M}^{\star}_{\text{FBM}}$ $(k \neq j)$, M_k is forbidden by the PI if it satisfies

$$\sum_{i \in \mathbb{N}_A} l_i \cdot M_k(p_i) \geq \sum_{i \in \mathbb{N}_A} l_i \cdot M_j(p_i) \tag{8.7}$$

where l_i's $(i \in \mathbb{N}_A)$ are the coefficients of I. By simplifying Eq. (8.7), we have

$$\sum_{i \in \mathbb{N}_A} l_i \cdot (M_k(p_i) - M_j(p_i)) \geq 0 \tag{8.8}$$

Similarly, we introduce a set of binary variables f_k's $(k \in \mathbb{N}^{\star}_{\text{FBM}}$ and $k \neq j)$ to represent the relation between I and any marking M_k in $\mathcal{M}^{\star}_{\text{FBM}}$. Then, Eq. (8.8) is modified as:

$$\sum_{i \in \mathbb{N}_A} l_i \cdot (M_k(p_i) - M_j(p_i)) \geq -Q \cdot (1 - f_k) \tag{8.9}$$

where Q is a positive integer constant that must be big enough and $f_k \in \{0, 1\}$. In Eq. (8.9), $f_k = 1$ indicates that M_k is forbidden by I and $f_k = 0$ indicates that it is not. All constraints for FBMs in $\mathcal{M}^{\star}_{\text{FBM}}$ are grouped as follows:

$$\sum_{i \in \mathbb{N}_A} l_i \cdot (M_k(p_i) - M_j(p_i)) \geq -Q \cdot (1 - f_k), \forall M_k \in \mathcal{M}^{\star}_{\text{FBM}} \text{ and } k \neq j \tag{8.10}$$

Given $M_j \in \mathcal{M}_{FBM}^\star$, we have Eq. (8.10). In addition, the coefficients of I should satisfy the reachability conditions. Therefore, the following ILPP to design a PI for M_j is derived, which is denoted as the Maximal number of Forbidding FBM Problem 2 (MFFP2).

MFFP2:

$$\max \ f = \sum_{k \in \mathbb{N}_{FBM}^\star, k \neq j} f_k$$

subject to

$$\sum_{i \in \mathbb{N}_A} l_i \cdot (M_l(p_i) - M_j(p_i)) \leq -1, \ \forall M_l \in \mathcal{M}_L^\star \qquad (8.11)$$

$$\sum_{i \in \mathbb{N}_A} l_i \cdot (M_k(p_i) - M_j(p_i)) \geq -Q \cdot (1 - f_k), \ \forall M_k \in \mathcal{M}_{FBM}^\star \text{ and } k \neq j \quad (8.12)$$

$$l_i \in \{0, 1, 2, \ldots\}, \ \forall i \in \mathbb{N}_A$$

$$f_k \in \{0, 1\}, \ \forall k \in \mathbb{N}_{FBM}^\star \text{ and } k \neq j$$

The objective function f is used to maximize the number of FBMs that are forbidden by I. If MFFP2 for an FBM dose not have a solution, then there does not exist a maximally permissive PI.

Theorem 8.2 *If MFFP2 for an FBM has no solution, the FBM cannot be forbidden by a maximally permissive PI.*

Proof By contradiction, suppose that there exists a maximally permissive PI I that forbids an FBM M. Then, no legal marking in $\mathcal{M}_L^\star$ is forbidden by I and the coefficients of I satisfy the reachability condition, i.e., Eq. (8.11). For Eq. (8.12), it can be seen that it always has a solution (for example, $f_k = 0$, $k \in \mathbb{N}_{FBM}^\star$). Thus, MFFP2 for M has a solution, which contradicts the condition of the theorem. □

Now, we analyze the number of constraints and variables in MFFP2. The type of Eq. (8.11) has $|\mathcal{M}_L^\star|$ constraints. Since $k \neq j$, the number of constraints of type Eq. (8.12) is $|\mathcal{M}_{FBM}^\star| - 1$. Thus, the total number of the constraints in an MFFP2 is $|\mathcal{M}_{FBM}^\star| + |\mathcal{M}_L^\star| - 1$. The number of variables l_i's ($i \in \mathbb{N}_A$) is $|P_A|$. Since $k \in \mathbb{N}_{FBM}^\star$ and $k \neq j$, the number of variables f_k's is $|\mathcal{M}_{FBM}^\star| - 1$. Thus, MFFP2 has $|P_A| + |\mathcal{M}_{FBM}^\star| - 1$ variables in total.

In Chapter 7, we present an ILPP (called MCPP) to obtain a supervisor with the minimal number of control places. Table 8.1 compares MFFP1 and MFFP2 with MCPP for a Petri net model in terms of the number of constraints and variables. Table 8.1 shows that the two parameters in MFFP1 are slightly bigger than those in MFFP2, respectively. Thus, the computational overhead of MFFP2 is in theory a little smaller than that of MFFP1. This can be verified by the experimental results presented in Section 8.4. It can also be seen that the number of their constraints and variables is much smaller than that in MCPP. Thus, their computational cost is much less than that of MCPP.

Table 8.1 Comparison of constraints and variables among MFFP1, MFFP2, and MCPP

Parameters	MFFP1	MFFP2	MCPP
No. of constraints	$|\mathcal{M}^{\star}_{\text{FBM}}| + |\mathcal{M}^{\star}_{L}|$	$|\mathcal{M}^{\star}_{\text{FBM}}| + |\mathcal{M}^{\star}_{L}| - 1$	$|\mathcal{M}^{\star}_{\text{FBM}}| \cdot (|\mathcal{M}^{\star}_{L}| + 2|\mathcal{M}^{\star}_{\text{FBM}}| - 1)$
No. of variables	$|P_A| + |\mathcal{M}^{\star}_{\text{FBM}}| + 1$	$|P_A| + |\mathcal{M}^{\star}_{\text{FBM}}| - 1$	$|\mathcal{M}^{\star}_{\text{FBM}}| \cdot (|P_A| + |\mathcal{M}^{\star}_{\text{FBM}}|)$

8.3 DEADLOCK PREVENTION POLICY

This section presents two iterative deadlock prevention policies, each of which can obtain a maximally permissive liveness-enforcing supervisor with a small number of control places. First, we provide an algorithm by using MFFP1.

Algorithm 8.1 Deadlock Prevention Policy by Using MFFP1

Input: Petri net model (N, M_0) of an AMS with $N = (P^0 \cup P_A \cup P_R, T, F, W)$.

Output: A maximally permissive controlled Petri net system (N_1, M_1).

1) Compute the set of FBMs $\mathcal{M}_{\text{FBM}}$ and the set of legal markings $\mathcal{M}_L$ for (N, M_0).

2) Compute the minimal covered set of FBMs $\mathcal{M}^{\star}_{\text{FBM}}$ and the minimal covering set of legal markings $\mathcal{M}^{\star}_L$ for (N, M_0).

3) $V_M := \emptyset$. /*V_M is used to denote the set of control places to be computed.*/

4) **while** $\{\mathcal{M}^{\star}_{\text{FBM}} \neq \emptyset\}$ **do**

   Design MFFP1 by the method presented in Section 8.2.

   Solve MFFP1. Let l_i's $(i \in \mathbb{N}_A)$ and β be the solution if $f^* \neq 0$. Otherwise, exit, as the maximally permissive Petri net supervisor does not exist.

   Design a PI I and a control place p_c by the method presented in Section 5.2.1.

   $V_M := V_M \cup \{p_c\}$ and $\mathcal{M}^{\star}_{\text{FBM}} := \mathcal{M}^{\star}_{\text{FBM}} - F_I$. /*$F_I$ is defined in Chapter 5.*/
   endwhile

5) Add all control places in V_M to (N, M_0) and denote the resulting net system as (N_1, M_1).

6) Output (N_1, M_1).

7) End.

This algorithm can obtain a maximally permissive liveness-enforcing supervisor with a small number of control places if at each iteration, $f^* \neq 0$. First, the reachability graph of a Petri net model is computed, from which the sets of legal markings and FBMs are derived. By using the vector covering approach, we obtain the minimal covering set of legal markings $\mathcal{M}^{\star}_L$ and the minimal covered set of FBMs $\mathcal{M}^{\star}_{\text{FBM}}$, which are usually much smaller than the sets of legal markings and FBMs, respectively. Then, an iterative process is carried out. At each iteration, we design an ILPP MFFP1 to obtain a PI that can forbid as many FBMs as possible. Once a non-zero solution of MFFP1 is obtained, a PI and a control place are accordingly found. Then, we remove all the FBMs that are forbidden by the computed PI from the minimal covered set of FBMs. This process is carried out until the minimal covered set of FBMs is empty. Finally, we add all the computed control places to the Petri net model. Thus, all FBMs are forbidden and the controlled system cannot enter the deadlock-zone.

Theorem 8.3 *Algorithm 8.1 can obtain a maximally permissive supervisor for a Petri net model of an AMS if there exists an optimal solution $f^* \neq 0$ for MFFP1 at each iteration.*

Proof At each iteration, if there exists a non-zero solution that satisfies the reachability conditions, i.e., Eq. (8.5) in MFFP1, then no legal markings are forbidden and a maximally permissive PI is obtained. Only the FBMs that are forbidden by the computed PI are removed from the minimal covered set of FBMs. At the same time, the PI forbids at least one FBM since $f^* \neq 0$. Therefore, all markings in $\mathcal{M}^\star_{\text{FBM}}$ are forbidden when the iterative process terminates. According to Corollary 5.3, all FBMs are forbidden, implying that the controlled system is live since it can never enter the DZ. Finally, the supervisor with the control places derived from all maximally permissive PIs is maximally permissive.   □

Next, we develop an algorithm by using MFFP2. It is similar to Algorithm 8.1 except the iteration step, i.e., Step 4. Therefore, only the modified iteration step is provided.

Algorithm 8.2 Deadlock Prevention Policy by Using MFFP2
　　Steps 1-3 and 5-7 are the same as those in Algorithm 8.1
　　4) **while** $\{\mathcal{M}^\star_{\text{FBM}} \neq \emptyset\}$ **do**
　　　　$\forall M \in \mathcal{M}^\star_{\text{FBM}}$, design MFFP2 for M by the method in Section 8.2.
　　　　Solve MFFP2. Let l_i's ($i \in \mathbb{N}_A$) be the solution if it exists. Otherwise, exit, as the maximally permissive Petri net supervisor does not exist.
　　　　　Design a PI I and a control place p_c by the method presented in Section 5.2.1.
　　　　$V_M := V_M \cup \{p_c\}$ and $\mathcal{M}^\star_{\text{FBM}} := \mathcal{M}^\star_{\text{FBM}} - F_I$.
　　　endwhile

This algorithm can also obtain a maximally permissive liveness-enforcing supervisor with a small number of control places. Its difference from Algorithm 8.1 is at Step 4. In Algorithm 8.2, an FBM is singled out and MFFP2 is solved to forbid not only the selected FBM but also as many FBMs as possible. Once an optimal solution of MFFP2 is obtained, a PI and a control place are accordingly designed. Then, we remove all the FBMs that are forbidden by the computed PI from the minimal covered set of FBMs. This process is carried out until the minimal covered set of FBMs is empty.

Theorem 8.4 *Algorithm 8.2 can obtain a maximally permissive supervisor for a Petri net model of an AMS if there exists an optimal solution to MFFP2 at each iteration.*

Proof By Corollary 5.4, if there exists a solution that satisfies Eq. (8.11) at each iteration, then a maximally permissive PI is obtained and no legal markings are forbidden. Only the FBMs that are forbidden by the computed PI are removed from the minimal covered set of FBMs. Therefore, all markings in $\mathcal{M}^{\star}_{\text{FBM}}$ are forbidden when the iterative process terminates. According to Corollary 5.3, the controlled system is live since it can never enter the DZ. Thus, the supervisor derived from all maximally permissive PIs is maximally permissive. □

We discuss the two methods presented in this chapter and the one in Chapter 7. Compared with the method in Chapter 7, the computational time of the two methods reported in this chapter is much shorter since the number of constraints in MFFP1 or MFFP2 is greatly smaller than that in MCPP. Both can obtain a supervisor with a small number of control places but cannot guarantee the minimality of the supervisory structure. However, experimental results show that the number of control places obtained by the presented methods is equal to or very close to the minimal number obtained in Chapter 7. The number of constraints and variables in MFFP2 is slightly smaller than that in MFFP1. Thus, the computational time of Algorithm 8.2 is in theory shorter than that of Algorithm 8.1. However, at each iteration, MFFP1 is easier to design since it can be obtained by simply deleting the constraints for FBMs that are forbidden by the PI computed at the last iteration. In summary, Algorithm 8.1 is easy to use and straightforward while Algorithm 8.2 is more efficient.

The Petri net model of an AMS shown in Fig. 8.1 is used to demonstrate the two algorithms. In this Petri net model, there are 11 places and eight transitions. The places can be classified into three groups: idle place set $P^0 = \{p_1, p_8\}$, operation

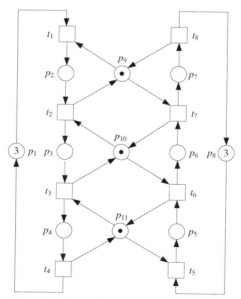

Figure 8.1 A simple Petri net model of an AMS.

place set $P_A = \{p_2, \ldots, p_7\}$, and resource place set $P_R = \{p_9, p_{10}, p_{11}\}$. The model has 20 reachable markings, 15 of which are legal and five are FBMs. By using the vector covering approach, the minimal covered set of FBMs and the minimal covering set of legal markings have three and two elements, respectively. Specifically, we have $\mathcal{M}_{FBM}^{\star} = \{p_2 + p_5, p_3 + p_5, p_2 + p_6\}$ and $\mathcal{M}_L^{\star} = \{p_2 + p_3 + p_4, p_5 + p_6 + p_7\}$.

First, Algorithm 8.1 is considered. At the first iteration, let I_1 be the PI to be computed. I_1 should satisfy the reachability conditions for the two legal markings in $\mathcal{M}_L^{\star}$, i.e., $l_2 \cdot 1 + l_3 \cdot 1 + l_4 \cdot 1 \leq \beta$ and $l_5 \cdot 1 + l_6 \cdot 1 + l_7 \cdot 1 \leq \beta$. Thus, we have the following two constraints

$$l_2 + l_3 + l_4 \leq \beta \text{ and}$$

$$l_5 + l_6 + l_7 \leq \beta.$$

We introduce three variables f_1, f_2, and f_3 $(f_1, f_2, f_3 \in \{0, 1\})$ to represent whether I_1 forbids $FBM_1 = p_2 + p_5$, $FBM_2 = p_3 + p_5$, and $FBM_3 = p_2 + p_6$, respectively. Thus, we have the following three constraints, i.e.,

$$l_2 + l_5 \geq \beta + 1 - Q \cdot (1 - f_1),$$

$$l_3 + l_5 \geq \beta + 1 - Q \cdot (1 - f_2), \text{ and}$$

$$l_2 + l_6 \geq \beta + 1 - Q \cdot (1 - f_3).$$

Finally, MFFP1 is defined as below.

MFFP1:

$$\max \quad f = f_1 + f_2 + f_3$$

subject to

$$l_2 + l_3 + l_4 \leq \beta$$

$$l_5 + l_6 + l_7 \leq \beta$$

$$l_2 + l_5 \geq \beta + 1 - Q \cdot (1 - f_1)$$

$$l_3 + l_5 \geq \beta + 1 - Q \cdot (1 - f_2)$$

$$l_2 + l_6 \geq \beta + 1 - Q \cdot (1 - f_3)$$

$$l_i \in \{0, 1, 2, \ldots\}, \ \forall i \in \{2, 3, 4, 5, 6, 7\}$$

$$\beta \in \{1, 2, \ldots\}$$

$$f_k \in \{0, 1\}, \ \forall k \in \{1, 2, 3\}$$

The above ILPP has an optimal solution with $l_2 = 2$, $l_5 = 1$, $l_6 = 1$, $\beta = 2$, $f_1 = 1$, $f_3 = 1$, and all the other variables are equal to zero. Then, a control place p_{c_1} is designed for I_1: $2\mu_2 + \mu_5 + \mu_6 + \mu_{p_{c_1}} = 2$. As a result, we have $M_0(p_{c_1}) = \beta = 2$, ${}^{\bullet}p_{c_1} = \{2t_2, t_7\}$ and $p_{c_1}^{\bullet} = \{2t_1, t_5\}$. FBM_1 and FBM_3 are forbidden by I_1, i.e., $F_{I_1} = \{FBM_1, FBM_3\}$. Thus, we have $\mathcal{M}_{FBM}^{\star} := \mathcal{M}_{FBM}^{\star} - F_{I_1}$, i.e., $\mathcal{M}_{FBM}^{\star} = \{p_3 + p_5\}$.

At the next iteration, let I_2 be the PI to be computed. The new MFFP1 can be easily obtained by deleting constraints with respect to FBM_1 and FBM_3 from the first MFFP1. As a result, we have

MFFP1:

$$\max \ f = f_2$$

subject to

$$l_2 + l_3 + l_4 \leq \beta$$

$$l_5 + l_6 + l_7 \leq \beta$$

$$l_3 + l_5 \geq \beta + 1 - Q \cdot (1 - f_2)$$

$$l_i \in \{0, 1, 2, \ldots\}, \ \forall i \in \{2, 3, 4, 5, 6, 7\}$$

$$\beta \in \{1, 2, \ldots\}$$

$$f_2 \in \{0, 1\}$$

The above ILPP has an optimal solution with $l_3 = 1$, $l_5 = 1$, $\beta = 1$, $f_2 = 1$, and all the other variables are equal to zero. Then, a control place p_{c_2} is designed for $I_2 : \mu_3 + \mu_5 + \mu_{p_{c_2}} = 1$. As a result, we have $M_0(p_{c_2}) = \beta = 1$, $^\bullet p_{c_2} = \{t_3, t_6\}$, and $p_{c_2}^\bullet = \{t_2, t_5\}$. Only FBM_2 is forbidden by I_2, i.e., $F_{I_2} = \{FBM_2\}$. Thus, we have $\mathcal{M}_{FBM}^\star := \mathcal{M}_{FBM}^\star - F_{I_2}$, i.e., $\mathcal{M}_{FBM}^\star = \emptyset$.

Now, Algorithm 8.1 terminates and there are totally two control places obtained for this net. Table 8.2 shows the results, where the first column is the iteration number i, the second shows the computed PI I_i, the third, denoted by $|F_{I_i}|$, is the number of FBMs in $\mathcal{M}_{FBM}^\star$, which are forbidden by I_i, the fourth to sixth columns indicate pre-transitions ($^\bullet p_{c_i}$), post-transitions ($p_{c_i}^\bullet$), and initial marking ($M_0(p_{c_i})$) of control place p_{c_i}, respectively, and the last two columns are the numbers of constraints and variables in each MFFP1, denoted by N_{LP} and N_{var}, respectively. Adding the two control places to the original net model, we obtain a live controlled system with 15 legal markings, as shown in Fig. 8.2. For this example, the MCPP in Chapter 7 has 18 constraints and 27 variables. Clearly, MFFP1 at the two iterations has much less constraints and variables (see N_{LP} in Table 8.2) and Algorithm 8.1 obtains the same result as in Chapter 7, i.e., a supervisor with the minimal number of control places.

Table 8.2 Control places computed for the net shown in Fig. 8.1 by Algorithm 8.1

i	I_i	$|F_{I_i}|$	$^\bullet p_{c_i}$	$p_{c_i}^\bullet$	$M_0(p_{c_i})$	N_{LP}	N_{var}
1	$2\mu_2 + \mu_5 + \mu_6 \leq 2$	2	$2t_2, t_7$	$2t_1, t_5$	2	5	10
2	$\mu_3 + \mu_5 \leq 1$	1	t_3, t_6	t_2, t_5	1	3	8

Now, we consider the application of Algorithm 8.2. At the first iteration, $FBM_1 = p_2 + p_5$ is selected. Let I_1 be the PI to forbid FBM_1, whose coefficients are l_i's ($i \in \{2, 3, 4, 5, 6, 7\}$). For the maximally permissive control purposes, I_1 should not forbid any marking in $\mathcal{M}_L^\star$. Thus, I_1 should satisfy two constraints: $l_2 \cdot (1 - 1) + l_3 \cdot (1 - 0) + l_4 \cdot (1 - 0) + l_5 \cdot (0 - 1) + l_6 \cdot (0 - 0) + l_7 \cdot (0 - 0) \leq -1$ and $l_2 \cdot (0 - 1) + l_3 \cdot (0 - 0) + l_4 \cdot (0 - 0) + l_5 \cdot (1 - 1) + l_6 \cdot (1 - 0) + l_7 \cdot (1 - 0) \leq -1$ for the two

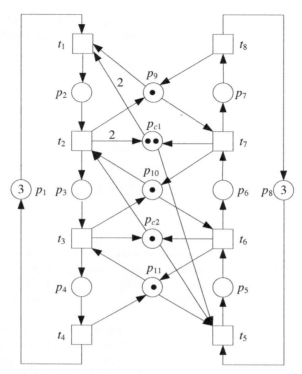

Figure 8.2 A controlled system of the Petri net in Fig. 8.1 by Algorithm 8.1.

legal markings $p_2 + p_3 + p_4$ and $p_5 + p_6 + p_7$ in $\mathcal{M}_L^\star$, respectively. By simplifying them, we have

$$l_3 + l_4 - l_5 \leq -1 \text{ and}$$

$$-l_2 + l_6 + l_7 \leq -1.$$

We introduce two variables f_2 and f_3 to represent whether I_1 forbids FBM$_2 = p_3 + p_5$ and FBM$_3 = p_2 + p_6$, respectively. Thus, we have the two constraints, i.e., $l_2 \cdot (0-1) + l_3 \cdot (1-0) + l_4 \cdot (0-0) + l_5 \cdot (1-1) + l_6 \cdot (0-0) + l_7 \cdot (0-0) \geq -Q \cdot (1-f_2)$ and $l_2 \cdot (1-1) + l_3 \cdot (0-0) + l_4 \cdot (0-0) + l_5 \cdot (0-1) + l_6 \cdot (1-0) + l_7 \cdot (0-0) \geq -Q \cdot (1-f_3)$, where Q is a positive integer constant that must be big enough and $f_2 = 1$ ($f_3 = 1$) indicates that FBM$_2$ (FBM$_3$) is forbidden by I_1 and $f_2 = 0$ ($f_3 = 0$) indicates that FBM$_2$ (FBM$_3$) cannot be forbidden by I_1. By simplifying them, we have

$$-l_2 + l_3 \geq -Q \cdot (1-f_2) \text{ and}$$

$$-l_5 + l_6 \geq -Q \cdot (1-f_3).$$

Finally, MFFP2 for FBM_1 is formulated as follows:

MFFP2:

$$\max \ f = f_2 + f_3$$

$$\text{subject to}$$
$$l_3 + l_4 - l_5 \leq -1$$
$$-l_2 + l_6 + l_7 \leq -1$$
$$-l_2 + l_3 \geq -Q \cdot (1 - f_2)$$
$$-l_5 + l_6 \geq -Q \cdot (1 - f_3)$$
$$l_i \in \{0, 1, 2, \ldots\}, \ \forall i \in \{2, 3, 4, 5, 6, 7\}$$
$$f_k \in \{0, 1\}, \ \forall k \in \{2, 3\}$$

The above ILPP has an optimal solution with $l_2 = 1$, $l_3 = 1$, $l_5 = 2$, $f_2 = 1$, and all the other variables being zero. Then, a control place p_{c_1} is designed for I_1: $\mu_1 + \mu_3 + 2\mu_5 + \mu_{p_{c_1}} = 2$. As a result, we have $M_0(p_{c_1}) = 2$, ${}^\bullet p_{c_1} = \{t_3, 2t_6\}$, and $p_{c_1}^\bullet = \{t_1, 2t_5\}$. FBM_1 and FBM_2 are forbidden by I_1, i.e., $F_{I_1} = \{FBM_1, FBM_2\}$. Thus, we have $\mathcal{M}_{FBM}^\star := \mathcal{M}_{FBM}^\star - F_{I_1}$, i.e., $\mathcal{M}_{FBM}^\star = \{p_2 + p_6\}$.

At the second iteration, we consider the last FBM, i.e., FBM_3. Let I_2 be the PI to forbid FBM_3. First, the coefficients of I_2 should satisfy the reachability conditions, i.e.,

$$l_3 + l_4 - l_6 \leq -1 \quad \text{and}$$
$$-l_2 + l_5 + l_7 \leq -1.$$

Since FBM_3 is the last element in $\mathcal{M}_{FBM}^\star$, no other FBMs in the set need to be forbidden by I_2. In this case, we solve the above integer linear system and obtain a solution with $l_2 = 1$, $l_6 = 1$, and all the other variables being zero. Then, a control place p_{c_2} is designed for I_2: $\mu_2 + \mu_6 + \mu_{p_{c_2}} = 2$. As a result, we have $M_0(p_{c_2}) = 1$, ${}^\bullet p_{c_2} = \{t_2, t_7\}$, and $p_{c_2}^\bullet = \{t_1, t_6\}$. Only FBM_3 is forbidden by I_2, i.e., $F_{I_2} = \{FBM_3\}$. Thus, we have $\mathcal{M}_{FBM}^\star := \mathcal{M}_{FBM}^\star - F_{I_2}$, i.e., $\mathcal{M}_{FBM}^\star = \emptyset$.

Now, Algorithm 8.2 terminates and there are totally two control places obtained for this net. Table 8.3 shows the results, where the selected FBM, denoted by FBM_i, is shown in the second column. Adding the two control places to the original net model, we obtain a live controlled net with 15 legal markings, as shown in Fig. 8.3. For this example, Algorithm 8.2 also obtains the same result as that in Chapter 7, i.e., a supervisor with the minimal number of control places.

Table 8.3 Control places computed for the net shown in Fig. 8.1 by Algorithm 8.2

i	FBM_i	I_i	$\|F_{I_i}\|$	${}^\bullet p_{c_i}$	$p_{c_i}^\bullet$	$M_0(p_{c_i})$	N_{LP}	N_{var}
1	$p_3 + p_5$	$\mu_2 + \mu_3 + 2\mu_5 \leq 2$	2	$t_3, 2t_6$	$t_1, 2t_5$	2	4	8
2	$p_2 + p_6$	$\mu_2 + \mu_6 \leq 1$	1	t_2, t_7	t_1, t_6	1	2	6

From this example, it can be seen that MFFP1 has more constraints and variables than MFFP2. However, MFFP1 at each iteration is similar to the ones at the previous iterations. It can be obtained by deleting the constraints for the FBMs only that are

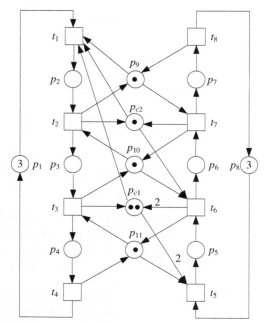

Figure 8.3 A controlled system of the Petri net in Fig. 8.1 by Algorithm 8.2.

forbidden by the PI computed at the last iteration. In summary, Algorithm 8.1 is easy to use and straightforward while Algorithm 8.2 is more efficient from the viewpoint of computational costs.

8.4 EXPERIMENTAL RESULTS

This section presents a number of AMS examples available in the literature to show the experimental results of the presented methods. The computational time refers to CPU's seconds in a computer under Windows XP operating system with Intel CPU Core 2.8 GHz and 4 GB memory.

The Petri net model of an AMS from (Uzam, 2002) is shown in Fig. 8.4. There are 16 places and 13 transitions. The places have the following set partitions: $P^0 = \{p_1, p_8\}$, $P_R = \{p_{13}, \ldots, p_{16}\}$, and $P_A = \{p_2, \ldots, p_7, p_9, \ldots, p_{12}\}$. It has 77 reachable markings, 13 and 64 of which are FBMs and legal ones, respectively. By using the vector covering approach, $\mathcal{M}_{\mathrm{FBM}}^{\star}$ and $\mathcal{M}_L^{\star}$ have five and 12 markings, respectively. We apply Algorithms 8.1 and 8.2 to the model, whose experimental results are shown in Tables 8.4 and 8.5, respectively, where the computational time of solving the ILPP at each iteration is shown in the last column, denoted by τ_{LP}.

For this example, the iterations in Algorithms 8.1 and 8.2 can be finished in 0.16s and 0.1s, respectively since the ILPP at each iteration has a smaller number of constraints and variables. However, for the method provided in Chapter 7, the

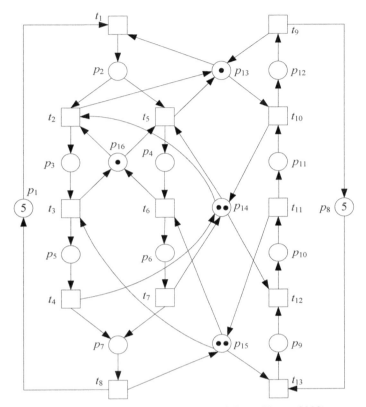

Figure 8.4 Petri net model of an AMS from (Uzam, 2002).

Table 8.4 Control places computed for the net shown in Fig. 8.4 by Algorithm 8.1

i	I_i	$|F_{I_i}|$	$^\bullet p_{c_i}$	$p_{c_i}^\bullet$	$M_0(p_{c_i})$	N_{LP}	N_{var}	$\tau_{LP}(S)$
1	$\mu_2 + \mu_3 + \mu_4 + 2\mu_9 \leq 4$	3	$t_3, t_6, 2t_{12}$	$t_1, 2t_{13}$	4	17	16	0.11
2	$\mu_2 + \mu_{10} + \mu_{11} \leq 1$	2	t_2, t_5, t_{10}	t_1, t_{12}	1	14	13	0.05

Table 8.5 Control places computed for the net shown in Fig. 8.4 by Algorithm 8.2

i	FBM_i	I_i	$|F_{I_i}|$	$^\bullet p_{c_i}$	$p_{c_i}^\bullet$	$M_0(p_{c_i})$	N_{LP}	N_{var}	$\tau_{LP}(S)$
1	$p_4 + 2p_9$	$\mu_2 + \mu_3 + \mu_4 + 2\mu_9 \leq 4$	3	$t_3, t_6, 2t_{12}$	$t_1, 2t_{13}$	4	16	14	0.05
2	$p_1 + p_{10}$	$\mu_2 + \mu_{10} + \mu_{11} \leq 1$	2	t_2, t_5, t_{10}	t_1, t_{12}	1	13	11	0.05

ILPP (denoted as MCPP in Algorithm 7.1) has 105 constraints and 75 variables. By solving the MCPP, we obtain an optimal solution in 0.17s with two control places only. Therefore, the two policies given in this chapter are slightly more efficient than Algorithm 7.1, and more importantly, can obtain a maximally permissive supervisor with two control places, i.e., the minimal number of control places.

Table 8.6 shows the available results for the example in terms of the numbers of the additional places and arcs, and the reachable markings of the controlled net. It can be seen that the results listed in the table are optimal but the presented method obtains a supervisor with two control places and 11 arcs only, where the number of the control places is the same as the one in Chapter 7 and only one more arc is generated.

Table 8.6 Performance comparison of typical deadlock control policies for the net in Fig. 8.4

Parameters	(Huang *et al.*, 2006)	(Uzam and Zhou, 2006)	Alg. 7.1	Alg. 8.1	Alg. 8.2
No. monitors	5	5	2	2	2
No. arcs	27	23	10	11	11
No. states	64	64	64	64	64

The Petri net model of an AMS is shown in Fig. 8.5, which has been studied in several papers (Uzam, 2002; Li *et al.*, 2008; Piroddi *et al.*, 2008, 2009). There are

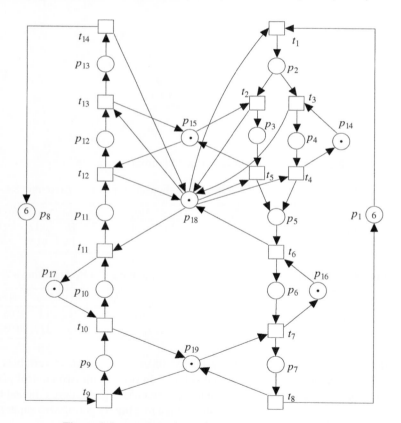

Figure 8.5 Petri net model of an AMS from (Uzam, 2002).

19 places and 14 transitions. The places have the following set partitions: $P^0 = \{p_1, p_8\}$, $P_R = \{p_{14}, \ldots, p_{19}\}$, and $P_A = \{p_2, \ldots, p_7, p_9, \ldots, p_{13}\}$. It has 282 reachable markings, 54 and 205 of which are FBMs and legal ones, respectively. By using the vector covering approach, $\mathcal{M}_{FBM}^\star$ and $\mathcal{M}_L^\star$ have eight and 26 markings, respectively. The applications of Algorithms 8.1 and 8.2 are shown in Tables 8.7 and 8.8, respectively.

Table 8.7 Control places computed for the net shown in Fig. 8.5 by Algorithm 8.1

i	I_i	$|F_{I_i}|$	$^\bullet p_{c_i}$	$p_{c_i}^\bullet$	$M_0(p_{c_i})$	N_{LP}	N_{var}	$\tau_{LP}(S)$
1	$2\mu_2 + 8\mu_3 + 4\mu_4 + 5\mu_5 + \mu_9 + \mu_{10} + 8\mu_{11} + 7\mu_{12} \le 14$	5	$3t_5, 5t_6, t_{12}, 7t_{13}$	$4t_1, 4t_2, t_4, t_9, 7t_{11}$	14	34	20	<0.01
2	$\mu_2 + 2\mu_3 + \mu_4 + 2\mu_5 + 2\mu_6 + 3\mu_9 + 3\mu_{10} \le 9$	3	$2t_7, 3t_{11}$	$t_1, t_2, t_4, 3t_9$	9	29	15	<0.01

Table 8.8 Control places computed for the net shown in Fig. 8.5 by Algorithm 8.2

i	FBM$_i$	I_i	$|F_{I_i}|$	$^\bullet p_{c_i}$	$p_{c_i}^\bullet$	$M_0(p_{c_i})$	N_{LP}	N_{var}	$\tau_{LP}(S)$
1	$p_2 + p_3 + p_4$	$\mu_2 + 2\mu_3 + \mu_4 + 2\mu_{11} + 2\mu_{12} \le 3$	4	$t_4, 2t_5, 2t_{13}$	$t_1, t_2, 2t_{11}$	3	33	18	<0.01
2	$p_3 + p_5 + p_9 + p_{10}$	$\mu_2 + 2\mu_3 + \mu_4 + 2\mu_5 + 2\mu_6 + 3\mu_9 + 3\mu_{10} \le 9$	4	$2t_7, 3t_{11}$	$t_1, t_2, t_4, 3t_9$	9	29	14	<0.01

For this example, the iterations in both Algorithms 8.1 and 8.2 can be finished in less than 0.02s since the ILPP at each iteration has a tiny number of constraints and variables. However, for the method presented in Chapter 7, the ILPP (denoted as MCPP in Algorithm 7.1) has 328 constraints and 152 variables, which can be solved in 1.1s. Therefore, the two policies introduced in this chapter are more efficient than Algorithm 7.1, and more importantly, can obtain a maximally permissive supervisor with two control places, i.e., the minimal number of control places. This example is widely used in the literature. Table 8.9 shows some available results for the example in terms of the numbers of the additional places and arcs, and the reachable markings of the controlled net. Note that Algorithm 7.1 obtains a supervisor with two control places and 12 arcs. Algorithm 8.2 obtains the same result and Algorithm 8.1 leads to a supervisor with the same number of control places and three more arcs than Algorithm 7.1.

Table 8.9 Performance comparison of typical deadlock control policies for the net in Fig. 8.5

Parameters	(Uzam, 2002)	(Li et al., 2008)	(Piroddi et al., 2008)	Alg. 5.3	Alg. 7.1	Alg. 8.1	Alg. 8.2
No. of monitors	6	9	5	8	2	2	2
No. of arcs	32	42	23	37	12	15	12
No. of states	205	205	205	205	205	205	205

Next, the Petri net model of a larger AMS (Ezpeleta *et al.*, 1995) is considered, as shown in Fig. 8.6. There are 26 places and 20 transitions. The places have the following set partitions: $P^0 = \{p_1, p_5, p_{14}\}$, $P_R = \{p_{20}, \ldots, p_{26}\}$, and $P_A = \{p_2, \ldots, p_4, p_6, \ldots, p_{13}, p_{15}, \ldots, p_{19}\}$. It has 26,750 reachable markings, 4,211 and 21,581 of which are FBMs and legal ones, respectively. By using the vector covering approach, $\mathcal{M}_{\text{FBM}}^{\star}$ and $\mathcal{M}_L^{\star}$ have 34 and 393 markings, respectively. The applications of Algorithms 8.1 and 8.2 are shown in Tables 8.10 and 8.11, respectively.

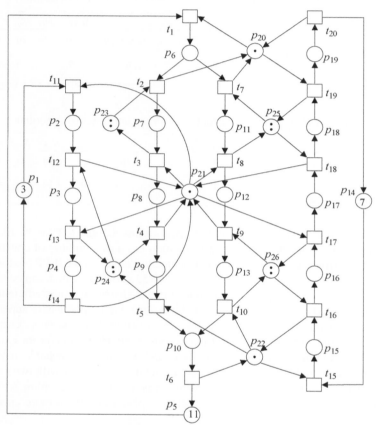

Figure 8.6 An S^3PR model from (Ezpeleta *et al.*, 1995).

Table 8.10 Control places computed for the net shown in Fig. 8.6 by Algorithm 8.1

i	I_i	$|F_{I_i}|$	$\bullet p_{c_i}$	$p_{c_i}^\bullet$	$M_0(p_{c_i})$	N_{LP}	N_{var}	$\tau_{LP}(S)$
1	$15\mu_2 + 15\mu_3 + 3\mu_6 + 3\mu_7 + 17\mu_8 + 17\mu_9 + 3\mu_{11} + 47\mu_{12} + 50\mu_{13} + 50\mu_{15} + 50\mu_{16} + \mu_{17} + 2\mu_{18} \le 196$	19	$17t_5, 50t_{10}, 15t_{13}, 49t_{17}, 2t_{19}$	$3t_1, 14t_3, 44t_8, 3t_9, 15t_{11}, 50t_{15}, t_{18}$	196	427	51	1.81
2	$\mu_2 + \mu_3 + \mu_8 + \mu_9 + 27\mu_{11} + 3\mu_{12} + 3\mu_{13} + 6\mu_{15} + 9\mu_{16} + 18\mu_{17} \le 71$	8	$t_5, 24t_8, 3t_{10}, t_{13}, 18t_{18}$	$t_3, 27t_7, t_{11}, 6t_{15}, 3t_{16}, 9t_{17}$	71	408	32	0.39
3	$6\mu_6 + 6\mu_7 + \mu_9 + 6\mu_{11} + \mu_{12} + \mu_{15} + \mu_{16} + 5\mu_{17} + 6\mu_{18} \le 34$	3	$6t_3, t_5, 5t_8, t_9, 6t_{19}$	$6t_1, t_4, t_{15}, 4t_{17}, t_{18}$	34	400	24	0.22
4	$\mu_2 + \mu_3 + \mu_8 \le 2$	2	t_4, t_{13}	t_3, t_{11}	2	397	21	0.05
5	$\mu_{12} + \mu_{16} \le 2$	1	t_9, t_{17}	t_8, t_{16}	2	395	19	0.06
6	$\mu_{13} + \mu_{15} \le 2$	1	t_{10}, t_{16}	t_9, t_{15}	2	394	18	0.05

Table 8.11 Control places computed for the net shown in Fig. 8.6 by Algorithm 8.2

i	FBM$_i$	I_i	$|F_{I_i}|$	$\bullet p_{c_i}$	$p_{c_i}^\bullet$	$M_0(p_{c_i})$	N_{LP}	N_{var}	$\tau_{LP}(S)$
1	$2p_{11} + 2p_{16}$	$8\mu_2 + 8\mu_3 + 2\mu_6 + 2\mu_7 + 9\mu_8 + 9\mu_9 + 75\mu_{11} + 25\mu_{12} + 25\mu_{13} + 50\mu_{15} + 75\mu_{16} + \mu_{17} + \mu_{18} \le 299$	10	$9t_5, 50t_8, 25t_{10}, 8t_{13}, 74t_{17}, t_{19}$	$2t_1, 7t_3, 73t_7, 8t_{11}, 50t_{15}, 25t_{16}$	299	426	49	0.26
2	$p_{12} + 2p_{16}$	$6\mu_2 + 6\mu_3 + \mu_6 + \mu_7 + 7\mu_8 + 7\mu_9 + \mu_{11} + 39\mu_{12} + 19\mu_{13} + 20\mu_{15} + 39\mu_{16} + \mu_{18} \le 116$	9	$7t_5, 20t_9, 19t_{10}, 6t_{13}, 20t_{15}, 19t_{16}, t_{18}$	$t_1, 6t_3, 38t_8, 6t_{11}, 20t_{15}, 19t_{16}, t_{18}$	116	416	39	0.33
3	$p_6 + 2p_7 + p_8 + p_9 + p_{11} + p_{13} + p_{15} + p_{16} + p_{18}$	$6\mu_2 + 6\mu_3 + \mu_6 + \mu_7 + 7\mu_8 + 7\mu_9 + \mu_{11} + 27\mu_{13} + 8\mu_{16} + \mu_{18} \le 80$	9	$7t_5, t_8, 27t_{10}, 6t_{13}, 19t_{16}, 8t_{17}, t_{19}$	$t_1, 6t_3, 27t_9, 6t_{11}, 27t_{15}, t_{18}$	80	407	30	0.22
4	$p_2 + 2p_3$	$\mu_2 + \mu_3 + \mu_8 \le 2$	2	t_4, t_{13}	t_3, t_{11}	2	398	21	0.05
5	$p_6 + 2p_7 + p_{11} + p_{17} + p_{18}$	$6\mu_6 + 6\mu_7 + \mu_9 + 23\mu_{11} + \mu_{13} + \mu_{15} + \mu_{16} + 23\mu_{17} + 5\mu_{18} \le 68$	3	$6t_3, t_5, 23t_8, t_{10}, 18t_{18}, 5t_{19}$	$6t_1, t_4, 17t_7, t_9, t_{15}, 22t_{17}$	68	396	19	0.11
6	$p_6 + 2p_7 + p_{17} + 2p_{18}$	$\mu_6 + \mu_7 + \mu_{17} + \mu_{18} \le 5$	1	t_3, t_7, t_{19}	t_1, t_{17}	5	393	16	<0.01

For this example, MCPP presented in Chapter 7 has 15,640 constraints and 1,700 variables. It requires about 44 hours to obtain an optimal solution. However, the iterations in Algorithms 8.1 and 8.2 can be finished within 2.58s and 0.98s, respectively. This example is also studied in a number of other papers (Ezpeleta *et al.*, 1995; Li and Zhou, 2004; Huang *et al.*, 2001; Piroddi *et al.*, 2008; Uzam and Zhou, 2006). Table 8.12 shows some typical available results for the example in terms of the numbers of the additional places and arcs, and the reachable markings of the controlled net. It can be seen that the results listed in the last five columns of the table are maximally permissive and Algorithm 7.1 obtains a supervisor with five control places (that are minimal for a maximally permissive supervisor) and 55 arcs. For this example, we conclude:

1. From the viewpoint of behavioral permissiveness, both Algorithms 8.1 and 8.2 can obtain an optimal supervisor.
2. As computational time is concerned, Algorithms 8.1 and 8.2 are efficient, reducing the CPU time for solving ILPPs from 44 hours by Algorithm 7.1 to 2.58s and 0.98s, respectively. The iterations in Algorithm 8.2 are finished in a shorter time than those in Algorithm 8.1.
3. About structural complexity, on the one hand, both Algorithms 8.1 and 8.2 obtain a supervisor with six control places, which has only one control place more than the minimal one obtained by Algorithm 7.1. On the other hand, Algorithm 8.1 obtains a supervisor with 45 arcs only, which is 10 less than the one computed by Algorithm 7.1. Algorithm 8.2 obtains a supervisor with 58 arcs, three more than those by Algorithm 7.1.

Table 8.12 Performance comparison of typical deadlock control policies for the net in Fig. 8.6

Parameters	(Li and Zhou, 2004)	(Huang *et al.*, 2001)	(Uzam and Zhou, 2006)	(Piroddi *et al.*, 2008)
No. of monitors	6	16	19	13
No. of arcs	32	88	112	82
No. of states	6287	12656	21562	21581

Parameters	Alg. 5.3	Alg. 7.1	Alg. 8.1	Alg. 8.2
No. of monitors	17	5	6	6
No. of arcs	101	55	45	58
No. of states	21581	21581	21581	21581

8.5 CONCLUSIONS

This chapter presents two iterative deadlock prevention policies to obtain a maximally permissive supervisor with a small number of control places. Both aim to overcome the computational bottleneck of the work introduced in the previous chapters. For example, the ILPP in Chapter 7 has too many constraints and variables for large Petri net models. Though the two methods are iterative, numerical studies show that they can terminate in a few iterations. Furthermore, the number of the constraints in the ILPP at each iteration is greatly lessened. Thus, the total computational time is also reduced. In the authors' opinion, the presented methods employ the idea of the divide-and-conquer strategy to tackle a complex problem.

The performance of a deadlock prevention policy is always evaluated by considering behavioral permissiveness, structural complexity, and computational complexity. For the two methods in this chapter, in terms of behavioral permissiveness, both are maximally permissive. From the aspect of structural complexity, it can reach or closely reach the smallest number of control places. As for the computational costs, the computational time is drastically reduced compared with that in Chapter 7.

The computational complexity of the considered problem is in theory NP-hard since both introduced methods require a complete enumeration of reachable markings and also need to solve ILPPs. Another problem is that the presented methods in theory cannot guarantee the minimality of the obtained supervisory structure.

8.6 BIBLIOGRAPHICAL REMARKS

Most materials of this chapter can be found in (Chen *et al.*, 2012). Compared with other work, for example, the behaviorally optimal methods in (Li *et al.*, 2008; Chen *et al.*, 2011; Ghaffari *et al.*, 2003; Uzam, 2002), the computationally efficient approaches in (Piroddi *et al.*, 2008, 2009; Huang *et al.*, 2001, 2006; Jeng and Xie, 2005), and structurally simple policies in (Chen and Li, 2011; Li and Zhou, 2004), this chapter provides a good trade-off among behavioral permissiveness, structural complexity, and computational overheads.

References

Banaszak, Z. and B. H. Krogh. 1990. Deadlock avoidance in flexible manufacturing systems with concurrently competing process flows. IEEE Transactions on Robotics and Automation. 6(6): 724–734.

Chen, Y. F., Z. W. Li, M. Khalgui, and O. Mosbahi. 2011. Design of a maximally permissive liveness-enforcing Petri net supervisor for flexible manufacturing systems. IEEE Transactions on Automation Science and Engineering. 8(2): 374–393.

Chen, Y. F. and Z. W. Li. 2011. An optimal liveness-enforcing supervisor with a compressed supervisory structure for flexible manufacturing systems. Automatica. 47(5): 1028–1034.

Chen, Y. F., Z. W. Li, and M. C. Zhou. 2012. Behaviorally optimal and structurally simple liveness-enforcing supervisors of flexible manufacturing systems. IEEE Transactions on Systems, Man, and Cybernetics, Part A. 42(3): 615–629.

Du, Y. Y., C. J. Jiang, and M. C. Zhou. 2009. A Petri net-based model for verification of obligations and accountability in cooperative systems. IEEE Transactions on Systems, Man, and Cybernetics, Part A. 39(2): 299–308.

Ezpeleta, J., J. M. Colom, and J. Martinez. 1995. A Petri net based deadlock prevention policy for flexible manufacturing systems. IEEE Transactions on Robotics and Automation. 11(2): 173–184.

Ezpeleta, J., F. Tricas, F. Garcia-Valles, and J. M. Colom. 2002. A banker's solution for deadlock avoidance in FMS with flexible routing and multiresource states. IEEE Transactions on Robotics and Automation. 18(4): 621–625.

Ghaffari, A., N. Rezg, and X. L. Xie. 2003. Design of a live and maximally permissive Petri net controller using the theory of regions. IEEE Transactions on Robotics and Automation. 19(1): 137–142.

Hsieh, F. S. and S. C. Chang. 1994. Dispatching-driven deadlock avoidance controller synthesis for flexible manufacturing systems. IEEE Transactions on Robotics and Automation. 10(2): 196–209.

Huang, Y. S., M. D. Jeng, X. L. Xie, and S. L. Chung. 2001. Deadlock prevention based on Petri nets and siphons. International Journal of Production Research. 39(2): 283–305.

Huang, Y. S., M. D. Jeng, X. L.Xie, and D. H. Chung. 2006. Siphon-based deadlock prevention for flexible manufacturing systems. IEEE Transactions on Systems, Man, and Cybernetics, Part A. 36(6): 1248–1256.

Jeng, M. D. and X. L. Xie. Deadlock detection and prevention of automated manufacturing systems using Petri nets and siphons. pp. 233–281. In M. C. Zhou and M. P. Fanti. [eds.]. 2005. Deadlock Resolution in Computer-Integrated Systems. New York: Marcel-Dekker Inc..

Lautenbach, K. and H. Ridder. 1996. The linear algebra of deadlock avoidance-a Petri net approach. No. 25–1996, Technical Report, Institute of Software Technology, University of Koblenz-Landau, Koblenz, Germany.

Li, Z. W. and M. C. Zhou. 2004. Elementary siphons of Petri nets and their application to deadlock prevention in flexible manufacturing systems. IEEE Transactions on Systems, Man, and Cybernetics, Part A. 34(1): 38–51.

Li, Z. W., M. C. Zhou, and M. D. Jeng. 2008. A maximally permissive deadlock prevention policy for FMS based on Petri net siphon control and the theory of regions. IEEE Transactions on Automation Science and Engineering. 5(1): 182–188.

Park, J. and S. A. Reveliotis. 2000. Algebraic synthesis of efficient deadlock avoidance policies for sequential resource allocation systems. IEEE Transactions on Robotics and Automation. 16(2): 190–195.

Park, J. and S. A. Reveliotis. 2001. Deadlock avoidance in sequential resource allocation systems with multiple resource acquisitions and flexible routings. IEEE Transactions on Automatic Control. 46(10): 1572–1583.

Piroddi, L., R. Cordone, and I. Fumagalli. 2008. Selective siphon control for deadlock prevention in Petri nets. IEEE Transactions on Systems, Man, and Cybernetics, Part A. 38(6): 1337–1348.

Piroddi, L., R. Cordone, and I. Fumagalli. 2009. Combined siphon and marking generation for deadlock prevention in Petri nets. IEEE Transactions on Systems, Man, and Cybernetics, Part A. 39(3): 650–661.

Tricas, F., F. Garcia-Valles, J. M. Colom, and J. Ezpelata. 1998. A structural approach to the problem of deadlock prevention in processes with resources, 273–278. *In* Proceedings of WODES'98, Italy, August 26–28.

Tricas, F., F. Garcia-Valles, J. M. Colom, and J. Ezpelata. An iterative method for deadlock prevention in FMS. pp. 139–148. *In* G. Stremersch [ed.]. 2000. Discrete Event Systems: Analysis and Control. Kluwer Academic: Boston, MA.

Uzam, M. 2002. An optimal deadlock prevention policy for flexible manufacturing systems using Petri net models with resources and the theory of regions. International Journal of Advanced Manufacturing Technology. 19(3): 192–208.

Uzam, M. and M. C. Zhou. 2006. An improved iterative synthesis method for liveness enforcing supervisors of flexible manufacturing systems. International Journal of Production Research. 44(10): 1987–2030.

Xing, K. Y., B. S. Hu, and H. X. Chen. Deadlock avoidance policy for flexible manufacturing systems. pp. 239–263. *In* M. C. Zhou. [ed.]. 1995. Petri Nets in Flexible and Agile Automation. Kluwer Academic: Boston, MA.

Chapter 9

Forbidden State Problems

ABSTRACT

This chapter deals with forbidden state problems by using Petri nets, which are a typical issue in supervisory control for DESs. Given a set of forbidden states in a plant net model, an optimal Petri net supervisor is developed. The optimality is twofold. On the one hand, it provides a minimal supervisory control structure in the sense of the number of monitors that are used to prevent the occurrences of the forbidden states in a plant model. A monitor is designed by associating a P-semiflow with other places in the plant. On the other hand, the supervisor is maximally permissive, i.e., no admissible state is excluded. If a maximally permissive Petri net supervisor does not exist, a most permissive supervisor is formulated. A net supervisor is said to be the most permissive if there are no other Petri net supervisors that are more permissive than it. A partial reachability graph of a plant is generated by considering the given forbidden states, from which FBMs are then identified. ILPPs are employed to offer monitor solutions to ensure that the FBMs are not reachable. The presented approaches are demonstrated through a number of examples.

9.1 INTRODUCTION

Given the discrete event model of a plant and its control specifications, a supervisory control problem is to synthesize a supervisor that couples with the plant such that the resulting system, called the controlled system, satisfies the specified control specifications. In most cases, a fundamental problem is to design a policy that guarantees that the system remains in a set of admissible states, or, equivalently, it never evolves into a specified set of forbidden states. Forbidden state problems are a typical class of control specifications in supervisory control theory of DESs, which are well formulated and addressed by Ramadge and Wonham in the framework of formal languages and automata (Ramadge and Wonham, 1987; Wonham and Ramadge, 1987; Ramadge and Wonham, 1989), namely R-W theory. However, R-W theory fails to make use of the structural characteristics of a plant, limiting it to be applied to the real-world systems.

As an alternative modeling formalism, Petri nets (Murata, 1989) have found their wide applications to modeling, analysis, control, and performance evaluation of

DESs (Girault and Valk, 2003; Iordache and Antsaklis, 2006; Reveliotis, 2005). Forbidden state problems (Holloway and Krogh, 1990) are extensively investigated in such a formalism (Dideban and Alla, 2005; Giua *et al.*, 1992; Iordache and Antsaklis, 2006).

The seminal work on feedback control logic synthesis is presented by Holloway and Krogh (Holloway and Krogh, 1990) for a class of Petri nets, called cyclic controlled marked graphs. Forbidden states cannot be accessed by establishing conditions that are used to disable controllable transitions.

As a particular case of forbidden state problems, the enforcement of generalized mutual exclusion constraints (GMECs) is studied in (Giua *et al.*, 1992) by using Petri nets with uncontrollable transitions. A GMEC indicates a condition that limits the weighted sum of tokens in a set of places. It is shown in (Giua *et al.*, 1992) that a forbidden state problem can be expressed as a GMEC for some particular classes of systems. A monitor (control place) is used to implement a GMEC by matrix operations.

The work in (Yamalidou *et al.*, 1996) extends the results in (Giua *et al.*, 1992) by using the concept of P-invariants of Petri nets. It considers logic and equality constraints and those involving firing vector elements and containing both marking and firing vector elements. By observing that the existing approaches usually lead to many monitors if the number of forbidden states is large, Dideban and Alla present a method to synthesize linear constraints given a set of forbidden states in a safe and conservative Petri net (Dideban and Alla, 2005). Given a set of forbidden states, their method aims to find a minimal number of linear constraints with respect to markings by using the P-invariants in a net. States are unreachable by forbidding the reachability of their over-state. The number of over-states is usually smaller than that of forbidden states. Then, an over-state finds a linear constraint. For each linear constraint, a monitor is calculated by the established technique in (Yamalidou *et al.*, 1996).

The study in (Dideban and Alla, 2008) proposes a method to reduce the number of constraints given a set of forbidden states. Border forbidden states are first identified from the reachability graph of a plant model and then their over-states are computed. The obtained over-states are reduced by considering the over-states of authorized (legal) states. Rules derived from the Quine-McCluskey algorithm, a method used for the minimization of boolean functions and developed by Quine and McCluskey, are used to find a minimal set of constraints. Each constraint is enforced by designing a monitor due to the technique in (Yamalidou *et al.*, 1996). Moreover, conditions are formulated under which there exists a maximally permissive supervisor. However, the method in (Dideban and Alla, 2008) can apply to safe Petri nets only and the number of monitors in a supervisor cannot be in general minimized.

Shortly, Vasiliu *et al.* (2009) extend the results in (Dideban and Alla, 2008) to non-safe Petri nets by modifying reduction rules and control conditions. The optimality of a supervisor is also considered, i.e., a maximally permissive supervisor can be definitely found if it exists in the form of Petri nets.

In the context of the aforementioned developments, and motivated by the above remarks, the study presented in this chapter makes the following key contributions.

1. First, forbidden state problems are considered using generalized Petri nets.
2. A maximally permissive monitor-based supervisor can be definitely found if it exists. If not, a most permissive monitor-based supervisor is provided such that there are no other Petri net supervisors more permissive than the most permissive one.
3. Structural complexity is also one of the most important criteria to evaluate the performance of a supervisor. This chapter provides a maximally or the most permissive monitor-based supervisor with a minimal structure (the number of monitors is minimized). In this sense, the chapter optimizes a supervisor behaviorally and structurally.

9.2 FORBIDDEN STATE PROBLEMS

Forbidden state problems are typical in supervisory control of DESs. In a Petri net formalism, M_f is usually used to denote the set of markings for which control specifications do not hold in a Petri net (N, M_0).

Definition 9.1 Let (N, M_0) be a Petri net and $M_f \subseteq R(N, M_0)$ a set of markings that do not satisfy a control specification. $M_F = M_f \cup \{M | M \in R(N, M_0); \exists \sigma \in T^*, \exists M' \in M_f, M'[\sigma\rangle M; \nexists \sigma' \in T^*, \forall M'' \in R(N, M_0) \setminus M_f, M''[\sigma'\rangle M\}$ is called the set of forbidden markings.

Property 9.1 If all markings in M_f are forbidden, none of markings in M_F is reachable.

The markings in M_F are hence unsafe. To find a Petri net supervisor for the given control specifications, the objective is to determine a set of monitors that, once added to a given plant net model, prevent the whole system from reaching the markings in M_F.

Definition 9.2 The set M_L of legal or admissible markings is the maximal set of reachable markings such that (1) $M_L \cap M_F = \emptyset$, and (2) it is possible to reach initial marking M_0 from any legal marking without leaving M_L.

By Definition 9.2, we have $M_L = R(N, M_0) \setminus M_F$. Let R_c be the reachability graph containing all legal markings only for given control specifications in a plant model (N, M_0). It is clear that every element in M_L can be found in R_c and every node in R_c is an element in M_L. At any marking in M_L, the system cannot be led outside M_L.

Figs. 9.1(a) and 9.1(b) show a Petri net (N, M_0) and its reachability graph $R(N, M_0)$, respectively. Suppose that a control specification is to prevent (N, M_0) from reaching markings M_3, M_4, and M_5. We have $M_f = \{M_3, M_4, M_5\}$, $M_F = \{M_3, M_4, M_5, M_6\}$, and $M_L = \{M_0, M_1, M_2, M_7, M_8, M_9\}$. R_c is hence the left part of the dashed line in Fig. 9.1(b). An optimal supervisor admits a set of monitors such that the reachability graph of the controlled system is exactly R_c. Fig. 9.1(c) depicts such an optimal supervisor for $\mathcal{M}_f = \{M_3, M_4, M_5\}$. To ensure the maximal permissiveness of a supervisor, $\mathcal{M}_L$ has to be found by computing a partial reachability graph for a plant.

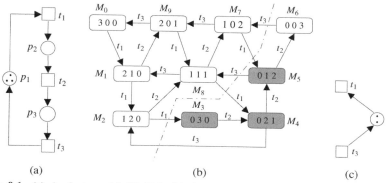

(a)   (b)   (c)

Figure 9.1 (a) A plant model (N, M_0), (b) the reachability graph of (N, M_0), and (c) an optimal supervisor.

9.2.1 Reachability Graph Generation under Forbidden States

A behaviorally optimal supervisor means that all the markings in R_c should be reachable in the controlled system and no markings in $\mathcal{M}_F$ are reachable. The following algorithm generates R_{cf}, a partial reachability graph that contains all the nodes in $\mathcal{M}_L \cup \mathcal{M}_f$.

Algorithm 9.1 Partial Reachability Graph Generation
  Input: Petri net (N, M_0) and $\mathcal{M}_f$.
  Output: R_{cf} and $\mathcal{M}_L$.
  1) The root node is M_0 that has initially no labels.
  2) **while** {there are nodes without labels} **do**
        Consider a node M without a label.
        foreach {transition t enabled at M} **do**
          $M' := M + [N](\cdot, t)$
          if {$M' \in \mathcal{M}_f$} **then**
            Label node M' "old".
          endif
          if {there does not exist a node M' in the graph} **then**
            Add it in R_{cf}.
            Add an arc t from M to M' in R_{cf}.
          endif
        Label node M "old".
      endwhile
  3) $\mathcal{M}_L := \{M | M \text{ is a node in } R_{cf}; M \notin \mathcal{M}_f\}$.
  4) End.

For the net shown in Fig. 9.1(a), Algorithm 9.1 does not generate M_6 since it is reachable from M_5 only and M_5 is a forbidden marking. Accordingly, $\mathcal{M}_L$ is the left part of the dashed line. The supervisor shown in Fig. 9.1(c) ensures that all the nodes in $\mathcal{M}_L$ are reachable and markings in $\mathcal{M}_f$ are not. This chapter presents an approach that finds a minimal set of monitors such that all the markings in $\mathcal{M}_L$ are reachable and all those in $\mathcal{M}_f$ are not.

In R_{cf}, the region that contains all the legal markings is called a legal zone, i.e., R_c, and the region that contains the markings in $\mathcal{M}_F$ is called a forbidden zone. The observation on the reachability graph generated by Algorithm 9.1 motivates us to conclude that, given a set of forbidden states, it is sufficient to forbid the reachability of a set of border nodes in the forbidden zone. Such nodes are called first-met bad markings (FBMs) that are proposed in (Uzam and Zhou, 2007) (see the definition in Section 2.2.3). Formally, we have:

Definition 9.3 The set of FBMs $\mathcal{M}_{FBM}$ in a plant (N, M_0) with $N = (P, T, F, W)$ is $\{M \in \mathcal{M}_F | \exists M' \in R_c, \exists t \in T, M'[t\rangle M\}$.

In Fig. 9.1(b), $\mathcal{M}_{FBM} = \{M_3, M_4, M_5\}$ since all markings in $\mathcal{M}_F$ can be forbidden if M_3, M_4, and M_5 are enforced to be not reachable.

9.2.2 Optimal Monitor Synthesis

This section introduces a method to obtain an optimal monitor, which is an important step in the design of an optimal supervisor to slove a forbidden state problem. In the following, $\mathbb{N}_n$ denotes $\{1, 2, \ldots, n\}$.

Definition 9.4 Let $\mathcal{M}_L$ and $\mathcal{M}_F$ be the sets of legal and forbidden markings in a plant (N, M_0), respectively. A monitor is said to be optimal or maximally permissive, if its addition to (N, M_0) forbids the reachability of one or more markings in $\mathcal{M}_F$ but all markings in $\mathcal{M}_L$ are still reachable.

Let μ be a reachable marking in a plant with n places and μ_i be the number of tokens in place p_i $(i \in \mathbb{N}_n)$ at μ. Any FBM $M \in \mathcal{M}_{FBM}$ can be forbidden by the following constraint:

$$\sum_{i \in \mathbb{N}_n} l_i \cdot \mu_i \leq \beta \tag{9.1}$$

where

$$\beta = \sum_{i \in \mathbb{N}_n} l_i \cdot M(p_i) - 1 \tag{9.2}$$

Eq. (9.1) is called the forbidding condition.

To ensure that every marking $M' \in \mathcal{M}_L$ cannot be prevented from being reached, coefficients l_i's $(i \in \mathbb{N}_n)$ in Eq. (9.1) should satisfy

$$\sum_{i \in \mathbb{N}_n} l_i \cdot M'(p_i) \leq \beta, \quad \forall M' \in \mathcal{M}_L \tag{9.3}$$

Eq. (9.3) is called the reachability conditions. For an FBM, by substituting Eq. (9.2) into Eq. (9.3), the reachability conditions of legal markings become

$$\sum_{i \in \mathbb{N}_n} l_i \cdot (M'(p_i) - M(p_i)) \leq -1, \quad \forall M' \in \mathcal{M}_L \qquad (9.4)$$

Eq. (9.4) determines coefficients l_i's ($i \in \mathbb{N}_n$). Thus, for an FBM M, if coefficients l_i's ($i \in \mathbb{N}_n$) of a PI satisfy Eq. (9.4), then the PI designed for Eq. (9.1) forbids M and none of the legal markings is forbidden. In this case, an optimal monitor can be designed by the method presented in Section 5.2.1 to ensure that all the legal markings can be reached.

9.2.3 Vector Covering Approach for a Minimal Supervisory Structure

In fact, it is unnecessary to deal with all FBMs. Next, we use a vector covering approach to reduce the number of FBMs that we need handle, which is similar to the vector covering technique shown in Chapter 5.

Definition 9.5 Let M and M' be two markings in $\mathcal{M}_{FBM}$. M covers M' (or M' is covered by M) if $\forall p \in P, M(p) \geq M'(p)$, which is denoted as $M \geq M'$ (or $M' \leq M$).

This definition is used to study the relations among different FBMs. Given this definition, we have the following theorem.

Theorem 9.1 Let M and M' be two markings in $\mathcal{M}_{FBM}$ with $M \geq M'$. If M' is forbidden by a PI, M is forbidden.

Proof Suppose that marking M' is forbidden by a PI that satisfies constraint $\sum_{i \in \mathbb{N}_n} l_i \cdot \mu_i \leq \beta$. In this case, we have $\beta = \sum_{i \in \mathbb{N}_n} l_i \cdot M'(p_i) - 1$. Since $M \geq M'$, we have $\forall i \in \mathbb{N}_n, M(p_i) \geq M'(p_i)$. It follows that $\sum_{i \in \mathbb{N}_n} l_i \cdot M(p_i) \geq \sum_{i \in \mathbb{N}_n} l_i \cdot M'(p_i) = \beta + 1$. This means that M is forbidden by the PI. □

Theorem 9.1 indicates that if a PI is designed to prevent a marking M' from being reached, then any marking M that covers M' is also unreachable. It is essential to reduce the set of the FBMs to be handled.

Definition 9.6 Let $\mathcal{M}_{FBM}^{\star}$ be a subset of $\mathcal{M}_{FBM}$. $\mathcal{M}_{FBM}^{\star}$ is called the minimal covered set of FBMs if the following two conditions are satisfied:

1. $\forall M \in \mathcal{M}_{FBM}, \exists M' \in \mathcal{M}_{FBM}^{\star}$, s.t. $M \geq M'$; and
2. $\forall M \in \mathcal{M}_{FBM}^{\star}, \nexists M'' \in \mathcal{M}_{FBM}^{\star}$, s.t. $M \geq M''$ and $M \neq M''$.

Corollary 9.1 If all markings of $\mathcal{M}_{FBM}^{\star}$ are forbidden by PIs, all the FBMs are forbidden.

Proof It follows immediately from Theorem 9.1. □

For a forbidden state problem, Corollary 9.1 indicates that only a small set of FBMs is considered to prevent all FBMs from being reached. Thus, the markings in $M_{FBM}^{\star}$ need to be considered only.

For an FBM $M \in M_{FBM}^{\star}$, it is unnecessary to consider all markings in M_L when designing an optimal monitor to forbid it. In the following, we provide a technique to reduce M_L to be a smaller set for forbidding M and the optimality of the monitor is still guaranteed.

Theorem 9.2 *For an FBM $M \in M_{FBM}^{\star}$, if $\forall M' \in M_L$, there exist coefficients l_i's ($i \in \mathbb{N}_n$) that satisfy Eq. (9.4), then a PI designed for Eq. (9.1) with $l_i^{\star}$ instead of l_i can also guarantee the reachability of all markings in M_L, where $\forall i \in \mathbb{N}_n$:*

$$l_i^{\star} = \begin{cases} l_i & \text{if } M(p_i) \neq 0 \\ 0 & \text{if } M(p_i) = 0 \end{cases} \tag{9.5}$$

Proof For an FBM $M \in M_{FBM}^{\star}$, according to the definition of $l_i^{\star}$'s ($i \in \mathbb{N}_n$), $\forall M' \in M_L$, we have:

$\sum_{i \in \mathbb{N}_n} l_i^{\star} \cdot (M'(p_i) - M(p_i))$

$= \sum_{i \in \mathbb{N}_n, M(p_i) \neq 0} l_i^{\star} \cdot (M'(p_i) - M(p_i)) + \sum_{i \in \mathbb{N}_n, M(p_i) = 0} l_i^{\star} \cdot (M'(p_i) - M(p_i))$

$= \sum_{i \in \mathbb{N}_n, M(p_i) \neq 0} l_i \cdot (M'(p_i) - M(p_i))$

$\leq \sum_{i \in \mathbb{N}_n, M(p_i) \neq 0} l_i \cdot (M'(p_i) - M(p_i)) + \sum_{i \in \mathbb{N}_n, M(p_i) = 0} l_i \cdot M'(p_i)$

$= \sum_{i \in \mathbb{N}_n, M(p_i) \neq 0} l_i \cdot (M'(p_i) - M(p_i)) + \sum_{i \in \mathbb{N}_n, M(p_i) = 0} l_i \cdot (M'(p_i) - M(p_i))$

$= \sum_{i \in \mathbb{N}_n} l_i \cdot (M'(p_i) - M(p_i))$

Since $\sum_{i \in \mathbb{N}_n} l_i \cdot (M'(p_i) - M(p_i)) \leq -1$, we have $\sum_{i \in \mathbb{N}_n} l_i^{\star} \cdot (M'(p_i) - M(p_i)) \leq -1$. This means that coefficients $l_i^{\star}$'s ($i \in \mathbb{N}_n$) can also satisfy the reachability conditions, i.e., Eq. (9.4), and lead to a PI that can guarantee the reachability of all markings in M_L. $\square$

This theorem indicates that an optimal monitor for an FBM can be designed by considering the places that are marked at the FBM only. Furthermore, a vector covering approach is developed to reduce the number of markings considered in M_L.

Definition 9.7 Let M_1 and M_2 be two markings in M_L. For an FBM $M \in M_{FBM}^{\star}$, M_1 M-covers M_2 (or M_2 is M-covered by M_1) if $\forall p \in P$ with $M(p) \neq 0$, $M_1(p) \geq M_2(p)$ is true, which is denoted as $M_1 \geq_M M_2$ (or $M_2 \leq_M M_1$).

Theorem 9.3 *For an FBM $M \in M_{FBM}^{\star}$, let M_1 and M_2 be two markings in M_L with $M_1 \geq_M M_2$. If M_1 is not forbidden by a PI designed for M, neither is M_2.*

Proof Suppose that M is forbidden by a PI that satisfies constraint $\sum_{i \in \mathbb{N}_n} l_i \cdot \mu_i \leq \beta$ with $\beta = \sum_{i \in \mathbb{N}_n} l_i \cdot M(p_i) - 1$. According to Theorem 9.2, the constraint can be reduced to be $\sum_{i \in \mathbb{N}_n, M(p_i) \neq 0} l_i \cdot \mu_i \leq \beta$ with $\beta = \sum_{i \in \mathbb{N}_n, M(p_i) \neq 0} l_i \cdot M(p_i) - 1$. In order not to forbid M_1, coefficients l_i's ($i \in \mathbb{N}_n, M(p_i) \neq 0$) of the PI should satisfy

$\sum_{i\in\mathbb{N}_n,M(p_i)\neq0} l_i \cdot M_1(p_i) \leq \beta$. Since $M_1 \geq_M M_2$, we have $\sum_{i\in\mathbb{N}_n,M(p_i)\neq0} l_i \cdot M_2(p_i) \leq \sum_{i\in\mathbb{N}_n,M(p_i)\neq0} l_i \cdot M_1(p_i) \leq \beta$. This means that M_2 is neither forbidden by the PI. □

Definition 9.8 Let $\mathcal{L}_M$ be a subset of $\mathcal{M}_L$. $\forall M \in \mathcal{M}^\star_{\text{FBM}}$, $\mathcal{L}_M$ is called the minimal covering set of M-related legal markings if the following two conditions are satisfied:

1. $\forall M' \in \mathcal{M}_L$, $\exists M'' \in \mathcal{L}_M$, s.t. $M'' \geq_M M'$; and
2. $\forall M' \in \mathcal{L}_M$, $\nexists M'' \in \mathcal{L}_M$, s.t. $M'' \geq_M M'$ and $M'' \neq M'$.

Corollary 9.2 *For an FBM $M \in \mathcal{M}^\star_{\text{FBM}}$, if M is forbidden by a PI and no element in the minimal covering set of M-related legal markings is forbidden, then none of the legal markings is forbidden.*

Proof It follows immediately from Theorem 9.3. □

Corollary 9.2 means that an optimal monitor for an FBM $M \in \mathcal{M}^\star_{\text{FBM}}$ can be designed to forbid M and ensures that no markings in the minimal covering set of its related legal markings $\mathcal{L}_M$ are forbidden. Therefore, for an FBM $M \in \mathcal{M}^\star_{\text{FBM}}$, Eq. (9.4) can be reduced to:

$$\sum_{i\in\mathbb{N}_n,M(p_i)\neq0} l_i \cdot (M'(p_i) - M(p_i)) \leq -1, \forall M' \in \mathcal{L}_M \qquad (9.6)$$

Usually, $\mathcal{L}_M$ is smaller than $\mathcal{M}_L$. Thus, the set of the inequalities for each FBM can be reduced.

9.3 STRUCTURALLY AND BEHAVIORALLY OPTIMAL SUPERVISOR

This section synthesizes a maximally permissive supervisor that minimizes the number of monitors. Such a supervisor is structurally and behaviorally optimal. In what follows, $\mathbb{N}^\star_{\text{FBM}}$ is used to denote $\{i|M_i \in \mathcal{M}^\star_{\text{FBM}}\}$.

For the minimality of a supervisory structure, all FBMs in $\mathcal{M}^\star_{\text{FBM}}$ must be considered. According to Theorem 9.2, only the places that are marked at FBMs are considered when we design optimal PIs. In this case, Given $\mathcal{M}^\star_{\text{FBM}}$, we define the support of $\mathcal{M}^\star_{\text{FBM}}$ as $\|\mathcal{M}^\star_{\text{FBM}}\| = \{p_i|\exists M \in \mathcal{M}^\star_{\text{FBM}}, M(p_i) > 0\}$. For the example in Fig. 9.1, we have $\mathcal{M}^\star_{\text{FBM}} = \mathcal{M}_{\text{FBM}} = \{M_3, M_4, M_5\}$ and $\|\mathcal{M}^\star_{\text{FBM}}\| = \{p_2, p_3\}$. Thus, only tokens in p_2 and p_3 are considered for this example. Next, we extend Definition 9.7 as follows.

Definition 9.9 Let M_1 and M_2 be two markings in $\mathcal{M}_L$. M_1 FBM-covers M_2 (or M_2 is FBM-covered by M_1) if $\forall p \in \|\mathcal{M}^\star_{\text{FBM}}\|$, we have $M_1(p) \geq M_2(p)$, which is denoted as $M_1 \geq_F M_2$ (or $M_2 \leq_F M_1$).

Theorem 9.4 *Let M_1 and M_2 be two markings in $\mathcal{M}_L$ with $M_1 \geq_F M_2$. If M_1 is not forbidden by a PI, neither is M_2.*

Proof It is similar to the proof of Theorem 9.3.   □

For the three markings M_0, M_1, and M_2 in Fig. 9.1(b), we have $M_2 \geq_F M_0$, $M_2 \geq_F M_1$, and $M_1 \geq_F M_0$. According to Theorem 9.4, if M_2 is not forbidden by any PI, neither are M_0 and M_1.

Definition 9.10 Let $\mathcal{L}_{FBM}$ be a subset of $\mathcal{M}_L$. $\mathcal{L}_{FBM}$ is called the minimal covering set of FBM-related legal markings if the following two conditions are satisfied:

1. $\forall M' \in \mathcal{M}_L$, $\exists M'' \in \mathcal{L}_{FBM}$, s.t. $M'' \geq_F M'$; and
2. $\forall M' \in \mathcal{L}_{FBM}$, $\nexists M'' \in \mathcal{L}_{FBM}$, s.t. $M'' \geq_F M'$ and $M'' \neq M'$.

Corollary 9.3 *If no markings in $\mathcal{L}_{FBM}$ are forbidden, then no legal markings are forbidden.*

Proof It follows immediately from Theorem 9.4.   □

Corollary 9.3 indicates that we consider $\mathcal{L}_{FBM}$ only for the optimal control purposes. For the net model shown in Fig. 9.1, we have $\mathcal{L}_{FBM} = \{M_2, M_7, M_8\}$. Finally, the reachability condition, i.e., Eq. (9.6), is modified as

$$\sum_{p_i \in \|M^\star_{FBM}\|} l_i \cdot (M'(p_i) - M(p_i)) \leq -1, \quad \forall M' \in \mathcal{L}_{FBM} \tag{9.7}$$

A PI may forbid multiple FBMs. Given a PI I_j for an FBM $M_j \in \mathcal{M}^\star_{FBM}$, any FBM $M_k \in \mathcal{M}^\star_{FBM}$ $(k \neq j)$ is forbidden if it satisfies the following constraints:

$$\sum_{p_i \in \|M_{FBM}\|} l_{j,i} \cdot M_k(p_i) \geq \sum_{p_i \in \|M_{FBM}\|} l_{j,i} \cdot M_j(p_i), \forall M_k \in \mathcal{M}^\star_{FBM} \text{ and } k \neq j \tag{9.8}$$

where $l_{j,i}$'s $(i \in \mathbb{N}_n)$ are the coefficients of I_j. By simplifying Eq. (9.8), we have

$$\sum_{p_i \in \|M_{FBM}\|} l_{j,i} \cdot (M_k(p_i) - M_j(p_i)) \geq 0, \forall M_k \in \mathcal{M}^\star_{FBM} \text{ and } k \neq j \tag{9.9}$$

For I_j, we introduce a set of variables $f_{j,k}$'s $(k \in \mathbb{N}_{FBM^\star}$ and $k \neq j)$ to represent the forbidding relationship between I_j and any marking M_k in $\mathcal{M}^\star_{FBM}$. Then, for each FBM in $\mathcal{M}^\star_{FBM}$, Eq. (9.9) is rewritten as:

$$\sum_{p_i \in \|M_{FBM}\|} l_{j,i} \cdot (M_k(p_i) - M_j(p_i)) \geq -Q \cdot (1 - f_{j,k}), \forall M_k \in \mathcal{M}^\star_{FBM} \text{ and } k \neq j \tag{9.10}$$

where Q is a positive integer constant that must be big enough and $f_{j,k} \in \{0, 1\}$. In Eq. (9.10), $f_{j,k} = 1$ indicates that M_k is forbidden by I_j and $f_{j,k} = 0$ indicates that M_k cannot be forbidden by I_j.

Now we introduce another set of variables h_j's $(j \in \mathbb{N}^\star_{FBM})$ for I_j, where $h_j = 1$ represents that I_j is selected to compute a monitor and $h_j = 0$ indicates that I_j is redundant and there is no need to add a monitor for it. If I_j is not selected, it cannot forbid any FBM. Therefore, we have the following conditions:

$$f_{j,k} \leq h_j, \quad \forall k \in \mathbb{N}_{\text{FBM}}^{\star} \text{ and } k \neq j \tag{9.11}$$

Any FBM M_j in $\mathcal{M}_{\text{FBM}}^{\star}$ must be forbidden by at least one PI. Thus, we have the following constraint:

$$h_j + \sum_{k \in \mathbb{N}_{\text{FBM}}^{\star}, \, k \neq j} f_{k,j} \geq 1 \tag{9.12}$$

For every FBM in $\mathcal{M}_{\text{FBM}}^{\star}$, we have the constraints represented by Eqs. (9.10), (9.11), and (9.12). Combining all these constraints and the reachability conditions together, we have the following ILPP to find a behaviorally and structurally optimal supervisor for a forbidden state problem, which is denoted by OSFS (Optimal Supervisor for Forbidden States).

OSFS:

$$\min h = \sum_{j \in \mathbb{N}_{\text{FBM}}^{\star}} h_j$$

subject to

$$\sum_{p_i \in \|\mathcal{M}_{\text{FBM}}^{\star}\|} l_{j,i} \cdot (M_l(p_i) - M_j(p_i)) \leq -1,$$

$$\forall M_j \in \mathcal{M}_{\text{FBM}}^{\star} \text{ and } \forall M_l \in \mathcal{L}_{\text{FBM}} \tag{9.13}$$

$$\sum_{p_i \in \|\mathcal{M}_{\text{FBM}}^{\star}\|} l_{j,i} \cdot (M_k(p_i) - M_j(p_i)) \geq -Q \cdot (1 - f_{j,k}),$$

$$\forall M_j, M_k \in \mathcal{M}_{\text{FBM}}^{\star} \text{ and } j \neq k \tag{9.14}$$

$$f_{j,k} \leq h_j, \ \forall j,k \in \mathbb{N}_{\text{FBM}}^{\star} \text{ and } j \neq k \tag{9.15}$$

$$h_j + \sum_{k \in \mathbb{N}_{\text{FBM}}^{\star}, \, k \neq j} f_{k,j} \geq 1, \ \forall j \in \mathbb{N}_{\text{FBM}}^{\star} \tag{9.16}$$

$$l_{j,i} \in \{0, 1, 2, \cdots\}, \ \forall i \in \mathbb{N}_n, p_i \in \|\mathcal{M}_{\text{FBM}}^{\star}\| \text{ and } \forall j \in \mathbb{N}_{\text{FBM}}^{\star}$$

$$f_{j,k} \in \{0, 1\}, \ \forall j,k \in \mathbb{N}_{\text{FBM}}^{\star} \text{ and } j \neq k$$

$$h_j \in \{0, 1\}, \ \forall j \in \mathbb{N}_{\text{FBM}}^{\star}$$

A feasible solution of this ILPP provides the minimal number of monitors such that each marking in $\mathcal{M}_F$ is forbidden and all the markings in $\mathcal{M}_L$ are reachable, as stated below.

Theorem 9.5 *A behaviorally optimal supervisor with the minimal number of control places can be obtained if OSFS has an optimal solution.*

Proof By Eq. (9.16), every FBM in $\mathcal{M}^{\star}_{\text{FBM}}$ is forbidden by at least one monitor in the final supervisor. Since all markings in $\mathcal{M}^{\star}_{\text{FBM}}$ are forbidden, then all FBMs are forbidden. According to Eq. (9.13), no legal markings are prohibited, i.e., all legal markings are reachable. From the objective function of OSFS, it is known that a minimal number of monitors are selected if an optimal solution is achieved. In a word, the final supervisor is both structurally and behaviorally optimal. □

Let us consider the forbidden state problem in Fig. 9.1(a), where $\mathcal{M}_{\text{FBM}} = \{M_3, M_4, M_5\}$ and $\mathcal{L}_{\text{FBM}} = \{M_2, M_7, M_8\}$. For this example, there is no covering relations among the three FBMs. Thus, $\mathcal{M}^{\star}_{\text{FBM}} = \mathcal{M}_{\text{FBM}}$. Let I_3 be the PI to forbid M_3, whose coefficients are $l_{3,i}$'s ($i \in \{2,3\}$). For the optimal control purposes, I_3 must not forbid any marking in $\mathcal{M}_L$. According to Eq. (9.7), we have

$$-l_{3,2} \leq -1,$$
$$-3l_{3,2} + 2l_{3,3} \leq -1, \text{ and}$$
$$-2l_{3,2} + l_{3,3} \leq -1.$$

We introduce two variables $f_{3,4}$ and $f_{3,5}$ ($f_{3,4}, f_{3,5} \in \{0,1\}$) to represent whether I_3 forbids M_4 and M_5, respectively. Thus, two constraints are found, i.e.,

$$-l_{3,2} + l_{3,3} \geq -Q \cdot (1 - f_{3,4}) \text{ and}$$
$$-2l_{3,2} + 2l_{3,3} \geq -Q \cdot (1 - f_{3,5})$$

where Q is a positive integer constant that must be big enough. For the former constraint, $f_{3,4} = 1$ indicates that M_4 is forbidden by I_3 and $f_{3,4} = 0$ indicates that this constraint is redundant and M_4 cannot be forbidden by I_3. Similarly, $f_{3,5} = 1$ indicates that M_5 is forbidden by I_3 and $f_{3,5} = 0$ indicates that this constraint is redundant and M_5 cannot be forbidden by I_3.

Similarly, for M_4, we have the following constraints:

$$-l_{4,3} \leq -1,$$
$$-2l_{4,2} + l_{4,3} \leq -1,$$
$$-l_{4,2} \leq -1,$$
$$l_{4,2} - l_{4,3} \geq -Q \cdot (1 - f_{4,3}), \text{ and}$$
$$-l_{4,2} + l_{4,3} \geq -Q \cdot (1 - f_{4,5}).$$

Finally, for M_5, we have:

$$l_{5,2} - 2l_{5,3} \leq -1,$$
$$-l_{5,2} \leq -1,$$
$$-l_{5,3} \leq -1,$$
$$2l_{5,2} - 2l_{5,3} \geq -Q \cdot (1 - f_{5,3}), \text{ and}$$
$$l_{5,2} - l_{5,3} \geq -Q \cdot (1 - f_{5,4}).$$

Next, we introduce a set of variables h_j's ($j \in \{3,4,5\}$), where $h_j=1$ indicates that I_j is selected to compute a monitor and $h_j = 0$ indicates that I_j is not selected. M_3 must be forbidden by at least one PI. Thus, we have the following constraint:

$$h_3 + f_{4,3} + f_{5,3} \geq 1.$$

Similarly, for M_4 and M_5, we have:

$$h_4 + f_{3,4} + f_{5,4} \geq 1$$

and

$$h_5 + f_{3,5} + f_{4,5} \geq 1.$$

If I_3 is not selected to compute a monitor, i.e., $h_3 = 0$, it cannot forbid M_4 and M_5. Therefore,

$$f_{3,4} \leq h_3 \quad \text{and}$$
$$f_{3,5} \leq h_3.$$

Similarly, for I_4 and I_5, we have:

$$f_{4,3} \leq h_4 \quad \text{and}$$
$$f_{4,5} \leq h_4,$$

and

$$f_{5,3} \leq h_5 \quad \text{and}$$
$$f_{5,4} \leq h_5.$$

Given all the above constraints, an objective function is used to minimize the number of monitors to be computed, as shown below:

$$\min \quad h_3 + h_4 + h_5$$

Grouping all the constraints above and the objective function, OSFS is formulated as follows:

OSFS:

$$\min \quad h = h_3 + h_4 + h_5$$

subject to

$$-l_{3,2} \leq -1$$

$$-3l_{3,2} + 2l_{3,3} \leq -1$$

$$-2l_{3,2} + l_{3,3} \leq -1$$

$$-l_{3,2} + l_{3,3} \geq -Q \cdot (1 - f_{3,4})$$

$$-2l_{3,2} + 2l_{3,3} \geq -Q \cdot (1 - f_{3,5})$$

$$-l_{4,3} \leq -1$$

$$-2l_{4,2} + l_{4,3} \leq -1$$

$$-l_{4,2} \leq -1$$

$$l_{4,2} - l_{4,3} \geq -Q \cdot (1 - f_{4,3})$$

$$-l_{4,2} + l_{4,3} \geq -Q \cdot (1 - f_{4,5})$$

$$l_{5,2} - 2l_{5,3} \leq -1$$

$$-l_{5,2} \leq -1$$

$$-l_{5,3} \leq -1$$

$$2l_{5,2} - 2l_{5,3} \geq -Q \cdot (1 - f_{5,3})$$

$$l_{5,2} - l_{5,3} \geq -Q \cdot (1 - f_{5,4})$$

$$h_3 + f_{4,3} + f_{5,3} \geq 1$$

$$h_4 + f_{3,4} + f_{5,4} \geq 1$$

$$h_5 + f_{3,5} + f_{4,5} \geq 1$$

$$f_{3,4} \leq h_3$$

$$f_{3,5} \leq h_3$$

$$f_{4,3} \leq h_4$$

$$f_{4,5} \leq h_4$$

$$f_{5,3} \leq h_5$$

$$f_{5,4} \leq h_5$$

$$l_{j,i} \in \{0,1,2,\cdots\}, \ \forall i \in \{2,3\} \text{ and } \forall j \in \{3,4,5\}$$

$$f_{j,k} \in \{0,1\}, \ \forall j,k \in \{3,4,5\} \text{ and } j \neq k$$

$$h_j \in \{0,1\}, \ \forall j \in \{3,4,5\}$$

The above OSFS has an optimal solution as shown in Table 9.1. In the solution, since $h_3 = 1$, $h_4 = 0$, and $h_5 = 0$, I_3 is selected to compute a monitor and there is no need computing monitors for I_4 and I_5. Note that in this solution, though I_4 (I_5)

that is designed to forbid M_4 (M_5) is not selected, M_4 (M_5) is forbidden by I_3 since $f_{3,4} = 1$ ($f_{3,5} = 1$). In fact, we can substitute M_4 (M_5) into I_3: $\mu_2 + \mu_3 \leq 2$. Then, we have $M_4(p_2) + M_4(p_3) = 3 \nleq 2$ ($M_5(p_2) + M_5(p_3) = 3 \nleq 2$), which contradicts the constraint of I_3. Thus, M_4 and M_5 are forbidden by I_3 and there is no need computing monitors for I_4 and I_5. A monitor associated with I_3 is computed, as shown in Fig. 9.1(c). Thus, for this example, the three forbidden states can be forbidden by only one optimal monitor, i.e., the minimal supervisory structure.

Table 9.1 An optimal solution of OSFS

Variable	h_3	$f_{3,4}$	$f_{3,5}$	$l_{3,2}$	$l_{3,3}$	h_4	$f_{4,3}$	$f_{4,5}$	$l_{4,2}$	$l_{4,3}$	h_5	$f_{5,3}$	$f_{5,4}$	$l_{5,2}$	$l_{5,3}$
Value	1	1	1	1	1	0	0	0	1	1	0	0	0	1	1

9.4 MOST PERMISSIVE SUPERVISOR DESIGN

Given a set of forbidden states for a Petri net model, there does not necessarily exist a behaviorally optimal liveness-enforcing supervisor that is expressed by a set of monitors. Then, deciding how to find a most permissive net supervisor such that no others are more permissive than it is interesting. This section aims to compute such a supervisor if a Petri net model has no optimal supervisor that can be represented by monitors.

Definition 9.11 Let M_1 and M_2 be two markings in $\mathcal{M}_L$. For an FBM $M \in \mathcal{M}_{FBM}^\star$, M_1 M-equals M_2 if $\forall p \in P$ with $M(p) \neq 0$, we have $M_1(p) = M_2(p)$, which is denoted as $M_1 =_M M_2$.

Definition 9.12 For an FBM $M \in \mathcal{M}_{FBM}^\star$, $\forall M_1 \in \mathcal{L}_M$, the set of legal markings that M-equal M_1 is defined as $\Xi_{M,M_1} = \{M_2 | M_2 \in \mathcal{M}_L, M_2 =_M M_1\}$.

Definition 9.12 indicates that when an FBM M is forbidden, if we cannot guarantee the reachability of legal marking M_1 in $\mathcal{L}_M$, then all legal markings that M-equal M_1 are forbidden. Next, we introduce a set of binary variables $f_1, f_2, \cdots,$ and $f_{N_\mathcal{L}}$ to the reachability condition for every legal marking in $\mathcal{L}_M$, where $N_\mathcal{L}$ is the cardinality of set $\mathcal{L}_M$. Then, Eq. (9.6) becomes

$$\sum_{i \in \mathbb{N}_n, M(p_i) \neq 0} l_i \cdot (M_j(p_i) - M(p_i)) \leq Q \cdot f_j - 1, \quad \forall M_j \in \mathcal{L}_M \qquad (9.17)$$

where Q is a positive constant integer that must be big enough and $f_j \in \{0, 1\}$. In Eq. (9.17), $f_j = 0$ indicates that the reachability condition of its corresponding marking M_j in $\mathcal{L}_M$ is satisfied and $f_j = 1$ indicates that this constraint is redundant and the corresponding marking M_j in $\mathcal{L}_M$ is not ensured to be reachable. Let N_{M,M_j} be the total number of markings that M-equal M_j, i.e., $N_{M,M_j} = |\Xi_{M,M_j}|$. The following ILPP can be used to ensure that the number of legal markings to be reachable is as large as possible, which is denoted as BSFS (Best Supervisor for Forbidden States).

BSFS:

$$\min \sum_{M_j \in \mathcal{L}_M} N_{M,M_j} \cdot f_j$$

subject to

$$\sum_{i \in \mathbb{N}_n, M(p_i) \neq 0} l_i \cdot (M_j(p_i) - M(p_i)) \leq Q \cdot f_j - 1, \ \forall M_j \in \mathcal{L}_M \qquad (9.18)$$

$$l_i \in \{0, 1, 2, \cdots\}, \forall i \in \mathbb{N}_n \text{ and } M(p_i) \neq 0$$

$$f_j \in \{0, 1\}, \forall M_j \in \mathcal{L}_M$$

The objective function represents the minimal number of legal markings that cannot be reached. Solving this ILPP, we have a solution of f_j's. If $f_j = 1$ in the solution, it means that M_j cannot be reached if the monitor is added. Thus, we should remove all the legal markings that M-equal M_j and compute a new set $\mathcal{L}_M$. A new ILPP is designed to ensure all the markings in the new set $\mathcal{L}_M$ to be reachable. Therefore, a most permissive supervisor can be obtained.

Algorithm 9.2 Design of a Most Permissive Supervisor

Input: Petri net (N, M_0) and $\mathcal{M}_F$.
Output: A set of monitors V_M.

1) Compute the set of FBMs $\mathcal{M}_{FBM}$ and the set of legal markings $\mathcal{M}_L$ for (N, M_0).
2) Compute the minimal covered set of FBMs $\mathcal{M}_{FBM}^\star$.
3) $V_M := \emptyset$. /* V_M is used to denote the set of monitors to be computed.*/
4) **while** $\{\mathcal{M}_{FBM}^\star \neq \emptyset\}$ **do**
   $\forall M \in \mathcal{M}_{FBM}^\star$, compute the minimal covered set of M-related legal markings $\mathcal{L}_M$.

   Solve the set of integer linear inequalities represented by Eq. (9.6).

   while {the set of integer linear inequalities represented by Eq. (9.6) has no solution} **do**
       Solve the ILPP BSFS.
       foreach $\{f_j = 1\}$ **do**
         $\mathcal{M}_L := \mathcal{M}_L - \Xi_{M,M_j}$.
         Compute the minimal covering set of M-related legal markings $\mathcal{L}_M$.
       endwhile
   Let l_i's ($i \in \mathbb{N}_A, M(p_i) \neq 0$) be the solution.
   Design a PI I and a monitor p_c by the method presented in Section 5.2.1.
   $V_M := V_M \cup \{p_c\}$, $\mathcal{M}_{FBM}^\star := \mathcal{M}_{FBM}^\star - F_I$.
   endwhile
5) End.

The algorithm first uses the vector covering approach. Then, the set of FBMs to be considered is reduced to $\mathcal{M}_{FBM}^\star$. This can reduce the total number of FBMs that should be forbidden by PIs. For each FBM M, the vector covering approach is used to reduce $\mathcal{M}_L$ to $\mathcal{L}_M$. This step can reduce the number of inequalities for each PI.

For each $M \in \mathcal{M}_{FBM}^{\star}$, if the ILPP has no solution, we design BSFS by introducing variables f_j's for all markings in $\mathcal{L}_M$ to represent their reachability. The objective function is used to ensure as many legal markings to be reached as possible. If $f_j = 0$, which indicates that the corresponding marking M_j cannot be reached, we remove from $\mathcal{M}_L$ all the markings that M-equal M_j. In this case, a new set of legal markings is generated and the ILPP for the FBM is redesigned. If it has no solution, we design BSFS to remove the minimal number of legal markings and resolve the ILPP. This process is carried out until the ILPP has a solution. As a result, a monitor is designed to forbid the FBM and it can guarantee the reachability of as many legal markings as possible. Finally, a most permissive supervisor can be obtained.

9.5 EXAMPLES

A Petri net model shown in Fig. 9.2 is used to illustrate the presented approaches. There are nine places and eight transitions. The net has 39 reachable markings and its reachability graph is shown in Fig. 9.3. Suppose that $\mathcal{M}_f = \{M_5, M_9, M_{11}, M_{12}, M_{13}, M_{18}, M_{22}, M_{26}, M_{30}, M_{38}\}$. Then, there are 22 legal markings and 10 FBMs. Using the vector covering approach to the set of FBMs, we have $\mathcal{M}_{FBM}^{\star} = \{p_1 + p_4, p_2 + p_4, p_1 + p_5, 2p_4 + p_5 + p_6\}$. There is no optimal supervisor for this problem. In the following, we design a most permissive supervisor. For an FBM M, since the places marked at M are considered only, any marking $M_j \in \mathcal{L}_M$ is denoted as $\sum_{M_j(p_i) \neq 0} M_j(p_i)p_i$. Taking the net shown in Fig. 9.1 as an example, for FBM M_4, we have $\mathcal{L}_{M_4} = \{M_2, M_7, M_8\}$. Then, the three markings in $\mathcal{L}_{M_4}$ are denoted as $M_2 = p_2$, $M_7 = 2p_3$, and $M_8 = p_2 + p_3$.

Now, we consider the application of Algorithm 9.2. At the first iteration, $FBM_1 = p_1 + p_4$ is selected. We have $\mathcal{L}_{FBM_1} = \{2p_4, 2p_1\}$. In order to forbid FBM_1, a PI I_1 is designed to satisfy constraint: $l_1 \cdot \mu_1 + l_4 \cdot \mu_4 \leq l_1 \cdot 1 + l_4 \cdot 1 - 1$. For the optimal control purposes, I_1 must not forbid any marking in $\mathcal{L}_{FBM_1}$. Thus, I_1 has to satisfy Eq. (9.7): $l_1 \cdot (0-1) + l_4 \cdot (2-1) \leq -1$ and $l_1 \cdot (2-1) + l_4 \cdot (0-1) \leq -1$ for the two legal markings $M_{L1} = 2p_4$ and $M_{L2} = 2p_1$ in $\mathcal{L}_{FBM_1}$, respectively. By simplifying them, we have

$$-l_1 + l_4 \leq -1 \quad \text{and}$$

$$l_1 - l_4 \leq -1.$$

The above integer linear program has no solution. In this case, we introduce two binary variables f_1 and f_2 to represent the reachability of the two legal markings in $\mathcal{L}_{FBM_1}$. The number of legal markings that FBM_1-equal M_{L1} is three, i.e., $|\Xi_{FBM_1,M_{L1}}| = 3$. Similarly, we have $|\Xi_{FBM_1,M_{L2}}| = 4$. Then, BSFS is presented below.

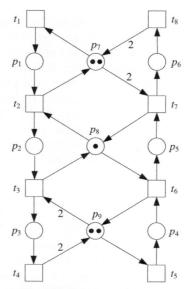

Figure 9.2 A Petri net example.

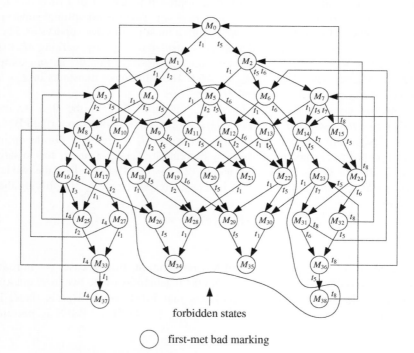

forbidden states

⃝ first-met bad marking

Figure 9.3 The reachability graph of the Petri net in Fig. 9.2.

BSFS:

$$\min \quad 3 \cdot f_1 + 4 \cdot f_2$$

subject to

$$-l_1 + l_4 \le Q \cdot f_1 - 1$$

$$l_1 - l_4 \le Q \cdot f_2 - 1$$

$$l_1, l_4 \in \{0, 1, 2, \cdots\}$$

$$f_1, f_2 \in \{0, 1\}$$

The above integer linear program has a solution with $f_1 = 1$ and $f_2 = 0$, where $f_1 = 1$ means that M_{L1} cannot be reached for the most permissive control purposes. Thus, all the markings that FBM_1-equal M_{L1} should be removed. After removing the three markings in $\Xi_{FBM_1, M_{L1}}$, the resulting set of legal markings has 19 elements. By using the vector covering approach, we have $\mathcal{L}_{FBM_1} = \{p_4, 2p_1\}$. In order to forbid FBM_1, I_1 is redesigned to satisfy constraint: $l_1 \cdot \mu_1 + l_4 \cdot \mu_4 \le l_1 \cdot 1 + l_4 \cdot 1 - 1$. For the most permissive control purposes, I_1 must not forbid any marking in $\mathcal{L}_{FBM_1}$. Thus, I_1 has to satisfy Eq. (9.6): $l_1 \cdot (0 - 1) + l_4 \cdot (1 - 1) \le -1$ and $l_1 \cdot (2 - 1) + l_4 \cdot (0 - 1) \le -1$ for the two legal markings p_4 and $2p_1$ in $\mathcal{L}_{FBM_1}$, respectively. By simplifying them, we have

$$-l_1 \le -1 \quad \text{and}$$

$$l_1 - l_4 \le -1.$$

The above integer linear program has a solution with $l_1 = 1$ and $l_4 = 2$. Then, a monitor p_{c_1} can be designed for I_1: $\mu_1 + 2\mu_4 + \mu_{p_{c_1}} = 2$ by the method presented in Section 5.2.1. As a result, we have $M_0(p_{c_1}) = 2$, $^\bullet p_{c_1} = \{t_2, 2t_6\}$, and $p_{c_1}^\bullet = \{t_1, 2t_5\}$. $FBM_1 = p_1 + p_4$ and $FBM_4 = 2p_4 + p_5 + p_6$ are forbidden by I_1, i.e., $F_{I_1} = \{FBM_1, FBM_4\}$. Hence, we have $\mathcal{M}_{FBM}^\star := \mathcal{M}_{FBM}^\star - \{FBM_1, FBM_4\}$, i.e., $\mathcal{M}_{FBM}^\star = \{p_2 + p_4, p_1 + p_5\}$.

At the second iteration, $FBM_2 = p_2 + p_4$ is selected. We have $\mathcal{L}_{FBM_2} = \{p_4, p_2\}$. In order to forbid FBM_2, a PI I_2 is designed to satisfy constraint: $l_2 \cdot \mu_2 + l_4 \cdot \mu_4 \le l_2 \cdot 1 + l_4 \cdot 1 - 1$. For the most permissive control purposes, I_2 must not forbid any markings in $\mathcal{L}_{FBM_2}$. Thus, I_2 has to satisfy Eq. (9.6): $l_2 \cdot (0 - 1) + l_4 \cdot (1 - 1) \le -1$ and $l_2 \cdot (1 - 1) + l_4 \cdot (0 - 1) \le -1$ for the two legal markings p_4 and p_2 in $\mathcal{L}_{FBM_2}$, respectively. By simplifying them, we have

$$-l_2 \le -1 \quad \text{and}$$

$$-l_4 \le -1.$$

The above integer linear program has a solution with $l_2 = 1$ and $l_4 = 1$. Then, a new monitor p_{c_2} can be designed for I_2: $\mu_2 + \mu_4 + \mu_{p_{c_2}} = 1$. As a result, we have $M_0(p_{c_2}) = 1$, $^\bullet p_{c_2} = \{t_3, t_6\}$, and $p_{c_2}^\bullet = \{t_2, t_5\}$. Only FBM_2 is forbidden by I_2, i.e., $F_{I_2} = \{FBM_2\}$. Hence, we have $\mathcal{M}_{FBM}^\star := \mathcal{M}_{FBM}^\star - \{FBM_2\}$, i.e., $\mathcal{M}_{FBM}^\star = \{p_1 + p_5\}$.

At the last iteration, FBM$_3$=$p_1 + p_5$ is selected. We have $\mathcal{L}_{\text{FBM}_3} = \{p_5, 2p_1\}$. In order to forbid FBM$_3$, a PI I_3 is designed to satisfy constraint: $l_1 \cdot \mu_1 + l_5 \cdot \mu_5 \le l_1 \cdot 1 + l_5 \cdot 1 - 1$. For the most permissive control purposes, I_3 must not forbid any marking in $\mathcal{L}_{\text{FBM}_3}$. Thus, I_3 also has to satisfy Eq. (5.13): $l_1 \cdot (0-1) + l_5 \cdot (1-1) \le -1$ and $l_1 \cdot (2-1) + l_5 \cdot (0-1) \le -1$ for the two legal markings p_5 and $2p_1$ in $\mathcal{L}_{\text{FBM}_3}$, respectively. By simplifying them, we have

$$-l_1 \le -1 \text{ and}$$

$$l_1 - l_5 \le -1$$

The above integer linear system has a solution with $l_1 = 1$ and $l_5 = 2$. Then, a new monitor p_{c_3} can be designed for I_3: $\mu_1 + 2\mu_5 + \mu_{p_{c_3}} = 2$. As a result, we have $M(p_{c_3}) = 2$, $^\bullet p_{c_3} = \{t_2, 2t_7\}$, and $p_{c_3}^\bullet = \{t_1, 2t_6\}$. Only FBM$_3$ is forbidden by I_3, i.e., $F_{I_3} = \{\text{FBM}_3\}$. Hence, we have $M_{\text{FBM}}^\star := M_{\text{FBM}}^\star - \{\text{FBM}_3\}$, i.e., $M_{\text{FBM}}^\star = \emptyset$.

Now, Algorithm 9.2 terminates and three monitors are obtained. Table 9.2 shows its application, where the first column is the iteration number i, the second represents the selected first-met bad marking FBM$_i$, and the third shows I_i for FBM$_i$. The fourth to sixth columns indicate pre-transitions ($^\bullet p_{c_i}$), post-transitions ($p_{c_i}^\bullet$), and initial marking ($M_0(p_{c_i})$) of monitor p_{c_i}, respectively. Adding the three monitors to the original net model, the resulting net has 19 legal markings. This is the most legal behavior by adding a set of monitors to the original model for the given forbidden state problem. That is to say, no other pure Petri net supervisors can be found, which can prohibit all the forbidden states and have more than 19 reachable states.

Table 9.2 Control places computed for the Petri net model in Fig. 9.2

i	FBM$_i$	I_i	$^\bullet p_{c_i}$	$p_{c_i}^\bullet$	$M_0(p_{c_i})$
1	$p_1 + p_4$	$\mu_1 + 2\mu_4 \le 2$	$t_2, 2t_6$	$t_1, 2t_5$	2
2	$p_2 + p_4$	$\mu_2 + \mu_4 \le 1$	t_3, t_6	t_2, t_5	1
3	$p_1 + p_5$	$\mu_1 + 2\mu_5 \le 2$	$t_2, 2t_7$	$t_1, 2t_6$	2

The structural complexity of the most permissive supervisor is not considered. In fact, we can use the method presented in Section 9.3 to design a most permissive supervisor with a minimal supervisory structure. An OSFS is designed to make the final 19 legal markings reachable and all the FBMs unreachable. Meanwhile, the objective function minimizes the number of the selected PIs. As a result, two PIs are obtained, i.e., a most permissive supervisor with the minimal structure consists of two monitors. The results are shown in Table 9.3.

Table 9.3 Best supervisor with a minimal structure for the net model in Fig. 9.2

i	FBM$_i$	I_i	$^\bullet p_{c_i}$	$p_{c_i}^\bullet$	$M_0(p_{c_i})$
2	$p_2 + p_4$	$\mu_1 + \mu_2 + 3\mu_4 \le 3$	$t_3, 3t_6$	$t_1, 3t_5$	3
3	$p_1 + p_5$	$\mu_1 + 2\mu_5 \le 2$	$t_2, 2t_7$	$t_1, 2t_6$	2

9.6 GMEC PROBLEMS

In supervisory control of DESs, forbidden state specifications are alternatively represented by generalized mutual exclusion constraints (GMECs). A GMEC is defined as a condition that limits a weighted sum of tokens contained in a subset of places (Giua *et al.*, 1992, 1993; Krogh and Holloway, 1991; Moody and Antsaklis, 1998) and includes both serial and parallel mutual exclusions (Zhou and DiCesare, 1993). Many constraints that deal with exclusions between states and events can be transformed into the form of GMECs (Iordache and Antsaklis, 2006).

Let $N = (P, T, F, W)$ be a net with initial marking M_0. A GMEC (l, b) to (N, M_0) is denoted by $(l, b) \equiv \sum_{p \in P} l(p) M(p) \leq b$, where (i) M is any reachable marking from M_0; (ii) $\forall p \in P$, $l(p)$ is a non-negative integer; and (iii) b is a non-negative integer. The following definitions can be found in (Giua *et al.*, 1992) and (Moody and Antsaklis, 1998).

Definition 9.13 Let (N, M_0) be a net system with place set P. A GMEC in N is defined as a set of legal markings $\mathcal{M}(l, b) = \{M \in \mathbb{N}^{|P|} | l^T M \leq b\}$, where l is called the characteristic P-vector and $b \in \mathbb{N}$ is called the constraint constant of the GMEC.

As defined previously, $\|l\| = \{p | l(p) \neq 0\}$ denotes the support of l. The markings in $\mathbb{N}^{|P|}$ that are not in $\mathcal{M}(l, b)$ are called forbidden markings with respect to constraint (l, b).

Definition 9.14 A set of GMECs (L, B) with $L = [l_1 | l_2 | \cdots | l_m]$ and $B = (b_1, b_2, \cdots, b_m)$ defines a set of legal markings $\mathcal{M}(L, B) = \{M \in \mathbb{N}^{|P|} | L^T M \leq B\} = \cap_{i=1}^{m} \mathcal{M}(l_i, b_i)$.

A GMEC can be enforced by a monitor (Giua *et al.*, 1992, 1993; Iordache and Antsaklis, 2006). Multiple GMECs need multiple monitors, leading to the structural complexity problem of a supervisor if there are too many GMECs in control specifications. Given a set of GMECs, a problem that naturally arises is to find a supervisor that minimizes the number of monitors, forbids the markings that do not satisfy the given set of GMECs, and ensures that all markings satisfying the given set of GMECs are reachable. To achieve these purposes, a computation method of a partial reachability graph under the GMECs is first presented.

Algorithm 9.3 Partial Reachability Graph Generation

Input: Petri net (N, M_0) and a set of GMECs (l_1, b_1), (l_2, b_2), …, and (l_m, b_m).
Output: R_{cf} and $\mathcal{M}_L$.

1) The root node is M_0 that has initially no labels.
2) **while** {there are nodes without labels} **do**
    Consider a node M without a label.
    foreach {transition t enabled at M} **do**
        $M' := M + [N](\cdot, t)$.
        flag1:=1
        for {$i = 1$ to m} **do**

if $\{M'$ does not satisfy $l_i^T M' \leq b_i\}$ **then**
    flag2:=0.
else
    flag2:=1.
endif
flag:=flag1×flag2.
endfor
if $\{$flag==0$\}$ **then**
    Label node M' "old".
endif
if $\{$there does not exist a node M' in the graph$\}$ **then**
    Add it in R_{cf}.
    Add an arc t from M to M' in R_{cf}.
endif
Label node M "old".
endwhile
3) $\mathcal{M}_L := \{M|M$ is a node in R_{cf}; M satisfies (l_1,b_1), (l_2,b_2), ..., and $(l_m,b_m)\}$.
4) End.

From R_{cf} and $\mathcal{M}_L$, the set of FBMs $\mathcal{M}_{\text{FBM}}$ can be accordingly obtained. Once $\mathcal{M}_L$ and $\mathcal{M}_{\text{FBM}}$ are available, a behaviorally and structurally optimal net supervisor can be computed if such a supervisor exists. Due to the limited space, we do not present the theoretical part for this issue, which is actually rather similar to what we have shown in the last sections. An example that is slightly modified from (Vasiliu *et al.*, 2009) is presented to demonstrate the approach.

Let us consider a manufacturing system with two machine tools M1 and M2, two robots Robot 1 and Robot 2, and three single-directed conveyors C1–C3. Its layout is shown in Fig. 9.4(a). Raw materials in C1(C2) can be uploaded to machine tool M1(M2) automatically. Finished parts are put in C3 and delivered. The processing capacity of each machine tool is two. Each robot can hold one part at a time. The machine tools are loaded automatically and once a part is finished on a machine tool, a robot is used to unload the machine tool. Fig. 9.4(b) depicts the Petri net model of the manufacturing system, where places p_1, p_4, and p_7 model M1, M2, and robots, respectively.

Suppose that there is a set of GMECs: $(l_1,b_1) \equiv p_2 + p_4 \leq 2$, $(l_2,b_2) \equiv p_4 + p_6 \leq 2$, $(l_3,b_3) \equiv p_2 + p_4 + p_6 \leq 2$, and $(l_4,b_4) \equiv 2p_1 + p_2 + p_4 \leq 6$. Under the GMECs, the partial reachability graph is shown in Fig. 9.5.

From the reachability graph, the nodes with gray background are forbidden markings and the others are legal. We have $\mathcal{M}_L = \{M_0, M_1, \ldots, M_{39}\}$ and $\mathcal{M}_F = \{M_{40}, M_{41}, \ldots, M_{53}\}$. To achieve the optimal control purposes, a minimal set of monitors have to be found to exclude the reachability of the forbidden states and to ensure that of all legal markings. By designing an OSFS and solving it, we can obtain an optimal solution. According to the solution, only one PI is selected. Specially, the selected PI is $\mu_2 + \mu_4 + \mu_6 \leq 2$ for FBM $(1\ 1\ 0\ 1\ 0\ 1\ 0\ 1)^T$. Thus, a monitor p_c is computed for the PI, where $M_0(p_c) = 0$, ${}^\bullet p_c = \{t_2, t_4\}$, and $p_c^\bullet = \{t_1, t_5\}$. Adding p_c

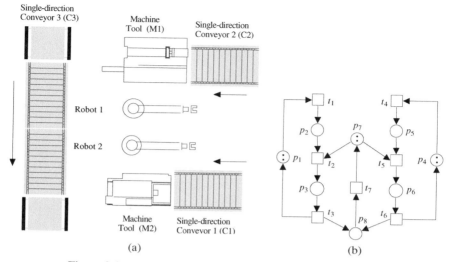

Figure 9.4 (a) The layout of an AMS and (b) its Petri net model.

to the original net model, all the 14 markings in $\mathcal{M}_F$ are forbidden and all elements in $\mathcal{M}_L$ are reachable. That is to say, all the given GMECs are enforced by only a monitor, i.e., the minimal supervisory structure.

9.7 CONCLUSIONS

This chapter addresses the issue of forbidden states, which is an important and typical control specification in supervisory control of DESs. Petri nets are used to model a system to be controlled. The existing approaches and their deficiencies in the literature motivate the development of the presented methods that provide an optimal Petri net supervisor, taking the form of monitors, in the sense of the permissive behavior and supervisory structure. The reported approach is applicable to all types of Petri nets. Although a full state space of a plant is avoided, this approach suffers from the computational complexity problem since a partial state enumeration is necessary.

9.8 BIBLIOGRAPHICAL REMARKS

Papers for forbidden state problems and GMECs can be found in (Dideban and Alla, 2005) and (Giua *et al.*, 1992, 1993; Li *et al.*, 2011), respectively. Optimization approaches for GMECs can be found in (Basile *et al.*, 2007; Dideban and Alla, 2008; Hu *et al.*, 2010).

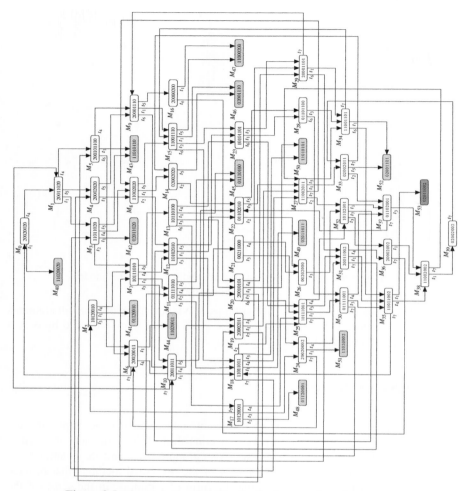

Figure 9.5 The partial reachability graph of a manufacturing system.

References

Basile, F., P. Chiacchio, and A. Guia. 2007. An optimization approach to Petri net monitor design. IEEE Transactions on Automatic Control. 52(2): 306–311.

Dideban, A. and H. Alla. 2005. From forbidden states to linear constraints for the optimal supervisor synthesis. Control Engineering and Applied Informatics. 7(3): 48–55.

Dideban, A. and H. Alla. 2008. Reduction of constraints for controller synthesis based on safe Petri nets. Automatica. 44(7): 1697–1706.

Girault, C. and R. Valk. 2003. Petri Nets for System Engineering: A Guide to Modeling, Verification, and Applications. New York: Springer-Verlag.

Giua, A., F. DiCesare, and M. Silva. 1992. Generalized mutual exclusion constraints on nets with uncontrollable transitions, 974–979. In Proceedings of IEEE International Conference on Systems, Man, and Cybernetics, Chicago, IL, U.S.A., Oct. 18–21.

Giua, A., F. DiCesare, and M. Silva. 1993. Petri net supervisor for generalized mutual exclusion constraints, 267–270. *In* Proceedings of the 12th IFAC World Congress, Sidney, Australia, July.

Holloway, L. E., and B. H. Krogh. 1990. Synthesis of feedback control logic for a class of controlled Petri nets. IEEE Transactions on Automatic Control. 35(5): 514–523.

Hu, H. S., M. C. Zhou, and Z. W. Li. 2010. Low-cost high-performance supervision in ratio-enforced automated manufacturing systems using timed Petri nets. IEEE Transactions on Automation Science and Engineering. 7(4): 933–944.

Li, Z. W., Y. F. Chen, and M. C. Zhou. 2011. On behaviorally and structurally optimal supervisor to solve forbidden state problems in discrete event systems, 832–837. IEEE Conference on Systems, Man, and Cybernetics, Alaska, October 9–12.

Iordache, M. V. and P. J. Antsaklis. 2006. Supervisory Control of Concurrent Systems: A Petri Net Structural Approach. Berlin: Springer-Verlag.

Iordache, M. V. and P. J. Antsaklis. 2006. Supervision based on place invariants: A survey. Discrete Event Dynamical Systems Theory and Applications. 16: 451–492.

Krogh, B. H. and L. E. Holloway. 1991. Synthesis of feedback control logic for discrete manufacturing systems. Automatica. 27(4): 641–651.

Moody, J. O. and P. J. Antsaklis. 1998. Supervisory Control of Discrete Event Systems Using Petri Nets. Norwell, MA: Kluwer.

Murata, T. 1989. Petri nets: Properties, analysis, and applications. Proceedings of the IEEE. 77(4): 541–580.

Ramadge, P. J. and W. M. Wonham. 1987. Supervisory control of a class of discrete-event processes. SIAM Journal Control Optimization. 25(1): 206–230.

Ramadge, P. J. and W. M. Wonham. 1989. The control of discrete event systems. Proceedings of the IEEE. 77(1): 81–98.

Reveliotis, S. A. 2005. Real-time Management of Resource Allocation Systems: A Discrete Event Systems Approach. NY: Springer-Verlag.

Uzam, M. and M. C. Zhou. 2007. An iterative synthesis approach to Petri net-based deadlock prevention policy for flexible manufacturing systems. IEEE Transactions on Systems, Man, and Cybernetics, Part A. 37(3): 362–371.

Vasiliu, A. I., A. Dideban, and H. Alla. 2009. Control synthesis for manufacturing systems using non-safe Petri nets. Control Engineering and Applied Informatics. 11(2): 43–50.

Wonham, W. M. and P. J. Ramadge. 1987. On the supremal controllable sublanguage of a given language. SIAM Journal of Control Optimization. 25(3): 637–659.

Yamalidou, E., J. O. Moody, and P. J. Antsaklis. 1996. Feedback control of Petri nets based on place invariants. Automatica. 32(1): 15–28.

Zhou, M. C. and F. DiCesare. 1993. Petri Net Synthesis for Discrete Event Control of Manufacturing Systems. Norwell, MA: Kluwer.

Chapter 10

Conclusions and Open Problems

ABSTRACT

This chapter first concludes the book by comparing all the presented deadlock prevention policies for AMSs in the sense of behavioral permissiveness, structural complexity, and computational complexity. Second, a number of interesting open problems in the development of deadlock prevention methods are provided such as the elementary siphons in more general Petri net classes, the uncontrollable and unobservable transitions, the implementation cost of an optimal supervisor, the design of more general (non-pure nets, self-loops and inhibitor arcs, for example) Petri net supervisors, the avoidance of generating all reachable markings, etc.

10.1 CONCLUSIONS

This particular monograph aims to present the state-of-the-art developments in the design of behaviorally and structurally optimal liveness-enforcing Petri net supervisors with computationally tractable approaches. The approaches published recently are included in this monograph, including the BDD-based computation and analysis of Petri nets, the theory of regions, the efficient computation of optimal supervisors, the design of the most permissive supervisors if no optimal one exists, the design of an optimal supervisors with the minimal number of control places, the design of behaviorally optimal and structurally simple supervisors, and the optimal supervisor design for forbidden state problems. The main novelties of the presented deadlock resolution are summarized below:

1. Presentation of a method to design a maximally permissive Petri net supervisor. Behavioral permissiveness plays an important role in evaluating the performance of supervisors for AMSs since an optimal, i.e., maximally permissive, supervisor always means high utilization of system resources. This book presents an effective and efficient method to design optimal Petri net supervisors, which can be used to deal with deadlock problems in large-scale AMSs. The concept of vectoring covering relations between markings is first provided in (Chen *et al.*, 2011) which can greatly reduce the computational overheads of the methods.

2. Presentation of an approach to design an optimal supervisor with a minimal supervisory structure. Structural complexity of a supervisor is also an important criterion in evaluating and designing a liveness-enforcing Petri net supervisor since a simple supervisory structure can decrease the hardware and software costs in the stage of control verification, validation, and implementation. In this book, we present a non-iterative method to design a behaviorally optimal and structurally minimal supervisor by solving an ILPP. Experimental results show that it can significantly reduce the structural complexity of an optimal supervisor. That is to say, it can lessen the implemental cost of optimal supervisors, which is both theoretically and practically important.

3. Presentation of a method to design a most permissive supervisor if no optimal one exists. There are systems that have no optimal pure Petri net supervisors. In this case, this book presents an approach to design a suboptimal, actually most permissive, supervisor for a system to be controlled. The resulting supervisor has the most number of legal states. Hence, we can ensure high utilization of system resources.

4. Presentation of a deadlock prevention method to find behaviorally optimal, structurally simple, and computationally efficient Petri net supervisors for AMSs. A deadlock control policy with low computational overheads means that it can be applied to large systems. This book presents deadlock prevention algorithms that can obtain liveness-enforcing supervisors with maximal permissiveness, a simple supervisory structure, and efficient computation.

5. Presentation of an efficient computation method of minimal siphons and symbolic analysis of the reachability graph of a Petri net by using BDDs. Minimal siphons and reachable markings are critical for the deadlock control in AMSs modeled by Petri nets. However, both increase exponentially with respect to the size of a Petri net. In order to make more efficient the deadlock prevention methods, BDDs are used in this book and a number of algorithms are shown to find minimal siphons and analyze the reachability graph of a Petri net. Experimental results show that BDDs are very powerful to deal with large sets of encoded data and make the methods applicable to large-scale Petri net models.

On the other hand, we do not completely solve the deadlock control problems in AMSs since there are still some shortcomings in the presented methods. First is the computational complexity. All methods are based on the analysis of reachability graphs of Petri nets, which always suffers from the state explosion problem. Meanwhile, the control places are computed by solving ILPPs that are NP-hard. Thus, the computational complexity is the main bottleneck of the methods.

In the following, we summarize all the presented methods in terms of behavioral permissiveness, structural complexity, and computational efficiency.

From the viewpoint of behavioral permissiveness, the methods in Chapters 4, 5, 7, and 8 can lead to an optimal Petri net supervisor if such a supervisor exists. In the case that there is no optimal supervisor for a Petri net model of an AMS, the approach in Chapter 6 can find a suboptimal, in fact the most permissive, liveness-enforcing supervisor.

From the viewpoint of structural complexity, the method in Chapter 7 can lead to an optimal supervisor with the minimal number of control places for Petri net models of AMSs. The one in Chapter 8 can lead to an optimal supervisor with a small number of control places, which is in fact very close to the minimal one as in Chapter 7. The approaches in Chapters 4 and 5 may lead to too many control places for a Petri net model in which redundant control places can survive.

In the sense of computational efficiency, the method in Chapter 5 can deal with large-scale Petri net models. The approach in Chapter 7 has to solve an ILPP with a large number of constraints and variables. Thus, it is time-consuming. However, it is the only non-iterative method such that all control places can be found by solving an ILPP once only. Chapter 8 presents a good trade-off between structural complexity and computational efficiency.

In summary, though a lot of work has been done for the deadlock control of AMSs, none can reach the optimization in all the three criteria: behavioral permissiveness, structural complexity, and computational complexity. In the authors' opinion, it still needs many efforts by researchers and engineers.

10.2 OPEN PROBLEMS

This section aims to outline the future directions of the development of deadlock control by using Petri nets and to facilitate readers in finding a suitable topic for their research.

10.2.1 Siphons in an S^3PR

10.2.1.1 Elementary siphon solution in an S^3PR

Elementary siphons (Li and Zhou, 2004, 2009) are shown to be an effective and efficient method to design structurally simple liveness-enforcing Petri net supervisors for AMSs. Their development is motivated by the observation that the controllability of some siphons can make a siphon controlled implicitly. The known results concerning elementary siphons are: (1) Their number in a Petri net is bounded by the smaller of place and transition counts and (2) The controllability of a dependent siphon can be ensured by explicitly controlling a set of elementary siphons. It can be used to develop deadlock prevention policies, leading to a structurally simple liveness-enforcing Petri net supervisor.

For a class of Petri nets, LS^3PR, an algorithm with polynomial complexity with respect to the net size is proposed in (Wang *et al.*, 2009). It is interesting for us to develop an algorithm with polynomial complexity, which can find a set of elementary siphons in an S^3PR that is more general than an LS^3PR. It is worthy of noting that elementary siphons in other manufacturing-oriented Petri net models that are more general than S^3PR are also interesting.

10.2.1.2 Structural condition of polynomial supremum of siphons

In theory, the number of the siphons in a Petri net grows quickly and in the worst case grows exponentially with respect to its size. In this sense, any deadlock prevention policy depending on a complete siphon enumeration is in theory of exponential complexity. Moreover, the structure of a liveness-enforcing supervisor suffers from the complexity problem since the number of monitors in a supervisor is in theory equal to that of minimal siphons that can be unmarked at a reachable marking.

In an LS^3PR, we find that the supremum of the strict minimal siphons is $2^n - n - 1$, where n is the number of resources that are a special class of places used to model machine tools, robots, and other manufacturing resources in a system.

We find that in most S^3PR, the number of strict minimal siphons that can be unmarked is actually not exponential with respect to the net size. However, we do find a structure of an S^3PR that leads to an exponential growth of siphons. A natural and appealing problem is to explore the structure of an S^3PR in which the number of strict minimal siphons is polynomial with respect to the net size. That is, if the supremum of siphons is polynomial with respect to the net size, siphon solution will not be exponential. Hence, for the class of Petri nets, we can develop deadlock prevention algorithms with polynomial complexity.

10.2.2 Iterative Deadlock Control Approach

Iterative control strategies are a naive but classical idea to deal with deadlock problems in an AMS, where its deadlocks are closely tied to the siphons in the Petri net model. The development of iterative deadlock control is motivated by the fact that the computation of a complete siphon enumeration is usually expensive. A recent result can be find in (Wang *et al.*, 2012).

10.2.2.1 Constringency

Siphon control in an ordinary Petri net is easier than that in a generalized case. That a siphon in an ordinary Petri net is not empty implies that a transition associated with it can fire at least once. This is also true in a PT-ordinary net. For the deadlock control purposes, we do not distinguish PT-ordinary and ordinary nets. Note that deadlock control in a generalized Petri net is much more difficult than that in an ordinary one, where the weight of an arc is one. This implies that the transitions in the postset of a marked siphon will not be disabled totally. That is to say, there necessarily exist enabled transitions in the postset of the marked siphon. Due to this, an elegant result in an ordinary net is developed, which is invariant-controlled siphons (Lautenbach and Ridder, 1993). A siphon is said to be invariant-controlled if it is a subset of the positive support of a P-invariant and the weighted token sum in the support of the invariant at an initial marking is greater than zero. An invariant-controlled siphon can never be unmarked at any reachable marking from the initial marking (Lautenbach and Ridder, 1993, 1996). However, the weight of an arc in a generalized Petri net can be an arbitrarily given positive integer such that it is difficult to properly decide the lower bound of the number of tokens in a siphon.

An iterative deadlock control approach is often concerned with net transformations and folding. There are two slightly different methods to perform net transformations and folding (Iordache et al., 2002; Lautenbach and Ridder, 1996). A net transformation means a generalized net is transformed into an ordinary one by adding extra places and transitions, called intermediate places and transitions, respectively. A net folding operation is to fold a net by removing the intermediate places and transitions and preserve the monitors added in the iteration processes. Net transformations, siphon computation and control, and net folding operations are typical steps in an iterative deadlock control approach. In the step of siphon computation and control, a complete siphon enumeration or a single siphon is usually found. A complete siphon enumeration in an iteration step is not recommended since its computation is time-consuming or even impossible.

An iterative deadlock control approach is usually considered to converge ultimately (Lautenbach and Ridder, 1996). Suppose that the reachability graph of a plant net model is finite. At each iteration step, one or more deadlock nodes are removed through the addition of monitors. Since the number of deadlock nodes is finite, the algorithm necessarily terminates at some step. The above statements look reasonable and are considered to be true in (Lautenbach and Ridder, 1996). However, in some cases, the iterative algorithm cannot terminate by using the net transformation method in (Lautenbach and Ridder, 1996), as shown in (Wang et al., 2012).

10.2.2.2 Behavioral optimality

Behavioral optimality of a supervisor derived from an iterative siphon control cannot be usually achieved. Even though a plant is ordinary, as the iteration steps proceed, the resulting net is prone to be generalized. As stated previously, it is difficult to decide the infimum of tokens in a siphon of a generalized Petri net. To fully eliminate the deadlocks, siphon control in a generalized Petri net is usually conservative (Barkaoui and Pradat-Peyre, 1996), leading to the fact that partial legal states are removed from the controlled system. A recent work is reported in (Piroddi et al., 2008, 2009), where siphon control and marking generation are combined. High behavioral permissiveness is achieved. Particularly, for three typical examples in the literature, behavioral optimality is achieved. However, in a general case, it remains open that how an optimal supervisor can be found via siphon control without a partial or complete marking enumeration.

10.2.3 Optimal Supervisor Design Problem

The performance of a liveness-enforcing supervisor can be evaluated by its computational complexity, structural complexity, and behavioral permissiveness. Many efforts are made to find a behaviorally optimal supervisor with a minimal supervisory structure and less computational costs.

10.2.3.1 Behaviorally optimal supervisor

The theory of regions that originally aims to provide a formal methodology to synthesize a Petri net from a transition system can be used to find a behaviorally optimal supervisor (Uzam, 2002). First, one generates the reachability graph of a plant Petri net model and then finds all MTSIs as well as the sets of legal and illegal markings. For an MTSI (M, t), a monitor is computed by solving an LPP such that its addition to the plant model disables t at M while ensures the reachability of all legal markings. The fatal disadvantage of the approaches based on the theory of regions is that a complete state enumeration is necessary. As known, the size of the reachability graph of a Petri net grows exponentially with respect to the number of its nodes and initial marking. This is the so-called state explosion problem.

Finding an MTSI can be done in polynomial or even linear time by a depth or breadth first search algorithm after a reachability graph is computed. However, for the deadlock control purposes of a Petri net, the number of MTSIs is in theory exponential with respect to the size of the model and its initial marking. Hence, the number of LPPs to be solved is in theory exponential with respect to the plant net size. In this sense, polynomial solvability of an LPP seems meaningless. Moreover, in such an LPP, the number of constraints is almost equal to that of the markings in a state space. Note that the number of LPPs to be solved in theory equals to that of MTSIs. However, a monitor can implement multiple MTSIs, leading to the fact that the number of monitors in a supervisor is generally much smaller than that of MTSIs in a reachability graph.

The major problem of the theory of regions is its computational complexity since a complete marking enumeration is a necessity. Finding such a supervisor without a complete state enumeration is an interesting issue. Moreover, deciding how to reduce the number of constraints in an LPP when a monitor is computed is also an important issue to decrease the computational overheads since in theory the number is close to the nodes in a reachability graph.

10.2.3.2 Structural complexity

Up to now, there is no conclusion that the number of monitors in a liveness-enforcing supervisor is bounded by the size of a plant net model. Finding a minimal supervisory structure is of significance. The work in (Chen and Li, 2011) computes a maximally permissive liveness-enforcing supervisor with a minimal supervisory structure where a monitor is associated with a P-semiflow. It remains open whether there exists a smaller structure in which a monitor is associated with a P-invariant, not a P-semiflow. Another issue is whether the number of monitors is definitely bounded by the size of a plant if the supervisory structure is minimal.

10.2.3.3 Uncontrollable and unobservable transitions

Uncontrollable and unobservable events in a plant may be present. Accordingly, it is reasonable and practical to consider their existence in a Petri net model of an FMS. Note that in RW-theory (Ramadge and Wonham, 1989), uncontrollable and unobservable events are sufficiently considered. However, Petri net researchers usually assume that all transitions are controllable and observable when a deadlock prevention policy is developed for an FMS. When the presence of uncontrollable and unobservable transitions is taken into account, most existing deadlock control policies need to be refined or even reinvestigated.

The work in (Qin *et al.*, 2011) considers the applicability of a deadlock prevention policy developed under the assumption that all transitions are controllable and observable, to a plant with uncontrollable and unobservable transitions. Sufficient and necessary conditions on uncontrollable and unobservable transitions have to be explored under which there exists a behaviorally and structurally optimal liveness-enforcing supervisor.

10.2.3.4 Non-pure optimal supervisor design

The work in (Chen *et al.*, 2012) considers the net models that cannot be optimally controlled by pure Petri net supervisors. However, there may exist non-pure Petri net supervisors (for example, including self-loops or inhibitor arcs) that can lead to optimally controlled systems. We know that self-loops and inhibitor arcs can greatly increase the modeling power of Petri nets. However, deciding how to mathematically represent a non-pure net structure is not an easy task.

10.2.4 Supervisor Design with Minimized Costs

An approach is established to find a behaviorally and structurally optimal supervisor with reasonable computational overheads in (Chen and Li, 2011; Chen *et al.*, 2011). An interesting problem is how to minimize the number of arcs in a behaviorally and structurally optimal supervisor. A more general case is to make each transition associate with a control cost. Deciding how to minimize the total control cost is also interesting. Another issue is to make minimal the sum of weights of the arcs from the monitors to transitions.

A WS^{3}PR can be live even if it has minimal siphons that do not contain marked traps (Zhong and Li, 2010). This fact is called self-liveness that is achieved by a proper marking and arc weights configuration. When a WS^3PR is not live, a reconfiguration of markings and arc weights is expected such that the resulting net system is live with a minimal regulation cost (Liu *et al.*, 2010).

10.2.5 Elementary Siphons in CPN or ROPN

ROPNs (Resource-Oriented PNs) are a compact modeling paradigm of AMS (Wu and Zhou, 2009). It is interesting to explore the theory of elementary siphons in ROPNs or CPNs (colored PNs) for the deadlock control purposes. The controllability of dependent siphons in an ROPN or CPN can be derived, which is similar to the results in (Li and Zhou, 2009).

10.2.6 Fault-tolerate Deadlock Control

The selection of deadlock control strategies depends on the frequency of deadlock occurrences in a system. If deadlocks are rather rare, a time-out mechanism may be accepted as the best approach to deal with deadlocks due to its low overhead. This is deadlock detection and recovery. In some cases, this strategy is not permitted due to technical or other factors. Instead, deadlocks are expected to forbid even if some resources break down. In a contemporary manufacturing system, automated equipment is widely and extensively used. The occurrences of faults in unreliable devices and machines can falsify a correctly designed deadlock prevention policy. Robust deadlock prevention and avoidance policies considering various errors and faults in an AMS are an interesting topic by using Petri nets as a formalism.

10.2.7 Existence of Optimal Supervisors

The existence of marking-based, not monitor-based, liveness-enforcing supervisors for discrete event systems is investigated by Sreenivas (Sreenivas, 1997a,b, 1999) in which the computation of a reachability graph is necessary. However, no sufficient attention is paid to the existence of an optimal (monitor-based) liveness-enforcing Petri net supervisor for an AMS. A natural and interesting problem is the structural and initial marking conditions of a Petri net under which there exists an optimal one. For example, whether there is an optimal supervisor for any S^3PR is interesting. For the existing manufacturing-oriented ordinary Petri net subclasses in the literature such as PPN and S^3PR, we have not seen any example whose optimal supervisors do not exist. However, it is easy to find an S^4PR whose optimal supervisors represented by pure Petri nets do not exist.

If the reachability graph of a Petri net has a maximal strongly connected component that contains the initial marking and all transitions are controllable, an optimal marking-based supervisor, i.e., an optimal supervisor taking the form of an automaton, exists. An interesting issue is the relationship between monitor- and marking-based supervisors enforcing liveness. For a bounded ordinary Petri net, it is shown in (He and Lemmon, 2000, 2002a,b) that (1) there exists a liveness-enforcing monitor if and only if there exists an optimal marking-based liveness-enforcing supervisor; and (2) a liveness-enforcing monitor solution may not be optimal. The results in (He and Lemmon, 2002a) are established by net unfolding techniques (McMillan, 1992, 1993) that map a Petri net to an acyclic occurrence net. A finite prefix of the occurrence net is defined to give a compact representation of the Petri net's reachability graph while preserving the causality between net transitions. This approach is used to deal with deadlock problems. A number of problems remain open. For example, it is appealing to find structural conditions under which an optimal monitor-based supervisor can be computed once an optimal marking-based supervisor exists.

10.2.8 Deadlock Avoidance with Polynomial Complexity

Different from deadlock prevention, deadlock avoidance is usually considered to be a technique that aims to check deadlock possibility dynamically and decides whether it is safe to grant a resource or not. It definitely needs extra information about the potential use of resources for each process. In a deadlock avoidance policy, the system dynamically considers every request and decides whether it is safe to grant it at the moment. The system requires additionally a priori information regarding the overall potential use of each resource for each process. In a general case, a deadlock avoidance problem is NP-hard.

A theoretically significant deadlock avoidance policy with polynomial-complexity is developed for a class of RASs in (Reveliotis et al., 1997), which is then described in a Petri net formalism (Park and Reveliotis, 2001). The work in (Xing et al., 2009) proposes an optimal deadlock avoidance policy with polynomial complexity for an S^3PR with a special initial marking, where a one-step look ahead method is used to check the safety of a reachable marking. Deciding the existence of an optimal deadlock avoidance policy for more general classes of Petri nets than an S^3PR is an interesting issue.

References

Barkaoui, K. and J. F. Pradat-Peyre. 1996. On liveness and controlled siphons in Petri nets, Lecture Notes in Computer Science. 1091: 57–72.

Chen, Y. F. and Z. W. Li. 2011. Design of a maximally permissive liveness-enforcing Petri net supervisor with a compressed supervisory structure for flexible manufacturing systems. Automatica. 47(5): 1028–1034.

Chen, Y. F., Z. W. Li, M. Khalgui, and O. Mosbahi. 2011. Design of a maximally permissive liveness-enforcing Petri net supervisor for flexible manufacturing systems. IEEE Transactions on Automation Science and Engineering. 8(2): 374–393.

Chen, Y. F., Z. W. Li, and M. C. Zhou. 2012. Most permissive liveness-enforcing Petri net supervisors for flexible manufacturing systems. International Journal of Production Research. 50(22): 6357–6371.

He, K. X. and M. D. Lemmon. 2000. Liveness verification of discrete-event systems modeled by n-safe ordinary Petri Nets. Lecture Notes in Computer Science. 1825: 227–243.

He, K. X. and L. D. Lemmon. 2002a. On the transformation of maximally permissive marking-based liveness enforcing supervisors into monitor supervisors, 2657–2662. *In* Proceedings of 39th IEEE Conference on Decision and Control, Sydney, NSW, Australia, December 12–15.

He, K. X. and M. D. Lemmon. 2002b. Liveness-enforcing supervision of bounded ordinary Petri nets using partial order methods. IEEE Transactions on Automatic Control. 47(7): 1042–1055.

Iordache, M. V., J. O. Moody, and P. J. Antsaklis. 2002. Synthesis of deadlock prevention supervisors using Petri nets. IEEE Transactions on Robotics and Automation. 18(1): 59–68.

Lautenbach, K. and H. Ridder. 1993. Liveness in bounded Petri nets which are covered by T-invariants. *In* Proceedings of the 13th International Conference on Applications and Theory of Petri Nets. Lecture Notes in Computer Science. 815: 358–375.

Lautenbach, K. and H. Ridder. 1996. The linear algebra of deadlock avoidance-a Petri net approach. No.25-1996, Technical Report, Institute of Software Technology, University of Koblenz-Landau, Koblenz, Germany.

Li, Z. W. and M. C. Zhou. 2004. Elementary siphons of Petri nets and their application to deadlock prevention in flexible manufacturing systems. IEEE Transactions on Systems, Man, and Cybernetics, Part A. 34(1): 38–51.

Li, Z. W. and M. C. Zhou. 2009. Deadlock Resolution in Automated Manufacturing Systems: A Novel Petri Net Approach. Springer, London, UK.

Liu, D., Z. W. Li, and M. C. Zhou. 2010. Liveness of an extended S^3PR. Automatica. 46(6): 1008–1018.

McMillan, K. 1992. Using unfoldings to avoid the state explosion problem in the verification of asynchronous circuits. Lecture Notes in Computer Science. 663: 164–177.

McMillan, K. 1993. Symbolic Model Checking. Norwell, Massachusetts: Kluwer.

Park, J. and S. A. Reveliotis. 2001. Deadlock avoidance in sequential resource allocation systems with multiple resource acquisitions and flexible routings. IEEE Transactions on Automatic Control. 46(10): 1572–1583.

Piroddi, L., R. Cordone, and I. Fumagalli. 2008. Selective siphon control for deadlock prevention in Petri nets. IEEE Transactions on Systems, Man, and Cybernetics, Part A. 38(6): 1337–1348.

Piroddi, L., R. Cordone, and I. Fumagalli. 2009. Combined siphon and marking generation for deadlock prevention in Petri nets. IEEE Transactions on Systems, Man, and Cybernetics, Part A. 39(3): 650–661.

Qin, M., Z. W. Li, M. Khalgui, and O. Mosbahi. 2011. On applicability of deadlock prevention policies with uncontrollable and unobservable transitions. International Journal of Innovative Computing, Information, and Control. 7(7B): 4115–4127.

Ramadge, P. J. and W. M. Wonham. 1989. The control of discrete event systems. Proceedings of the IEEE. 77(1): 81–89.

Reveliotis, S. A., M. A. Lawley, and P. M. Ferreira. 1997. Polynomial-complexity deadlock avoidance policies for sequential resource allocation systems. IEEE Transactions on Automatic Control. 42(10): 1344–1357.

Sreenivas, R. S. 1997a. On Commoner's liveness theorem and supervisory policies that enforce liveness in free-choice Petri nets. Systems Control Letters. 31(1): 41–48.

Sreenivas R. S. 1997b. On the existence of supervisory control policies that enforce liveness in discrete-event dynamic systems modeled by controlled Petri nets. IEEE Transactions on Automatic Control. 42(7): 928–945.

Sreenivas, R. S. 1999. On supervisory policies that enforce liveness in completely controlled Petri nets with directed cut-places and cut-transitions. IEEE Transactions on Automatic Control. 44(6): 1221–1225.

Uzam, M. 2002. An optimal deadlock prevention policy for flexible manufacturing systems using Petri net models with resources and the theory of regions. International Journal of Advanced Manufacturing Technology. 19(3): 192–208.

Wang, A. R., Z. W. Li, J. Y. Jian, and M. C. Zhou. 2009. An effective algorithm to find elementary siphons in a class of Petri nets. IEEE Transactions on Systems, Man, and Cybernetics, Part A. 39(4): 912–923.

Wang, A. R., Z. W. Li, M. C. Zhou, and A. Al-Ahmari. 2012. Iterative deadlock control by using Petri nets. IEEE Transactions on Systems, Man, and Cybernetics, Part C. DOI: 10.1109/TSMCC.2012.2189385.

Wu, N. Q. and M. C. Zhou. 2009. System Modeling and Control with Resource-Oriented Petri Nets. NY: CRC Press.

Xing, K. Y., M. C. Zhou, H. X. Liu, and F. Tian. 2009. Optimal Petri-net-based polynomial-complexity deadlock-avoidance policies for automated manufacturing systems. IEEE Transactions on Systems, Man, and Cybernetics, Part A. 39(1): 188–199.

Zhong, C. F. and Z. W. Li. 2010. On self-liveness of a class of Petri net models for flexible manufacturing systems. IET Control Theory and Applications. 4(3): 403–410.

Glossary

2^A	The power set of a set A
$\|A\|$	The number of elements in a set A
$B = \{0, 1\}$	A set called the carrier
C_L	The set of non-oriented cycles in the LZ
E_t	The enabled function of t
f	A function
F_I	The set of FBMs that are forbidden by a PI I
F_M	The set of FBMs that A-covers M
$F_n(B)$	A set of n-variables Boolean functions
$G(N, M_0))$	The reachability graph of net (N, M_0)
I	A place invariant
$\|I\|$	The support of a place invariant I
$\|I\|^+$	The positive support of a place invariant I
$\|I\|^-$	The negative support of a place invariant I
$\mathcal{I}$	An identity matrix
J	A transition invariant
M	A marking
$M(S)$	The sum of tokens in a place set S
$\mathcal{M}$	A set of markings
$\mathcal{M}_F$	A set of forbidding markings
$\mathcal{M}_L$	A set of legal markings
$\mathcal{M}_{\text{FBM}}$	A set of first-met bad markings
$\mathcal{M}_L^\star$	A minimal covering set of legal markings
$\mathcal{M}_{\text{FBM}}^\star$	A minimal covered set of first-met bad markings
$\mathcal{L}_{\text{FBM}}$	A minimal covering set of FBM-related legal markings
$\mathcal{L}_M$	A minimal covering set of M-related legal markings
(L, B)	A set of GMECs
$\mathcal{M}(L, B)$	A set of legal markings for (L, B)
N	A Petri net with $N = (P, T, F, W)$
N_A	The number of operation places
N_{FBM}	The number of first-met bad markings
N_L	The number of legal markings
$N_\mathcal{L}$	The number of elements in $\mathcal{L}_M$
N_{MS}	The number of minimal siphons
N_{M, M_j}	The number of markings that M-equal M_j
N_r	The number of reachable markings
N_S	The number of siphons
$N_{\text{FBM}}^{\text{BDD}}$	The number of BDD nodes for FBMs

N_l^{BDD}	The number of BDD nodes for legal markings
N_{MS}^{BDD}	The number of BDD nodes for minimal siphons
N_r^{BDD}	The number of BDD nodes for reachable markings
N_S^{BDD}	The number of BDD nodes for siphons
$\mathbb{N}$	The set of non-negative integers, $\mathbb{N} = \{0,1,2,\ldots\}$
$\mathbb{N}^+$	The set of positive integers, $\mathbb{N} = \{1,2,\ldots\}$
$\mathbb{N}_A$	$\{i\|p_i \in P_A\}$
$\mathbb{N}_L$	$\{i\|M_i \in \mathcal{M}_L\}$
$\mathbb{N}_{\text{FBM}}$	$\{i\|M_i \in \mathcal{M}_{\text{FBM}}\}$
$\mathbb{N}_L^\star$	$\{i\|M_i \in \mathcal{M}_L^\star\}$
$\mathbb{N}_{\text{FBM}}^\star$	$\{i\|M_i \in \mathcal{M}_{\text{FBM}}^\star\}$
$\mathbb{N}_m$	$\{1,2,\ldots,m\}$
(N, M_0)	A Petri net system
$[N]$	The incidence matrix
$[N]^+$	The output incidence matrix
$[N]^-$	The input incidence matrix
$[N_c]$	The incidence matrix of control places
n_c	The number of constraints
P	A set of places
P^0	A set of idle places
P_A	A set of operation (activity) places
P_R	A set of resource places
p	A place in a Petri net
$p^\bullet$	The postset of a place p
$^\bullet p$	The preset of a place p
p_0	An idle place
p_c	A control place, also called a monitor
Q	A big enough integer constant
$R(N, M_0)$	The set of reachable markings of (N, M_0)
$R(-N, M_0)$	The set of co-reachable markings of (N, M_0)
R_c	The reachability graph for given control specifications
R_{cf}	A partial reachability graph that contains all the nodes in $\mathcal{M}_L \cup \mathcal{M}_f$
R_M	The set of legal markings that is A-covered by M
$\mathcal{R}$	A relation
$\mathcal{R}_\subset$	A binary relation for $\subset$
r	A resource place
S	A siphon
$\mathcal{S}$	A set of siphons
$\mathcal{S}_{MS}$	The set of minimal siphons
$\mathcal{S}_N$	The set of siphons in N
T	A set of transitions
t	A transition in a Petri net
$^\bullet t$	The preset of a transition t
$t^\bullet$	The postset of a transition t

V_M	A set of control places
$^\bullet x$	The preset of a node $x \in P \cup T$
$x^\bullet$	The postset of a node $x \in P \cup T$
X	A set
$|X|$	The element count in a set X
$^\bullet X$	The preset of a set $X \subseteq P \cup T$
$X^\bullet$	The postset of a set $X \subseteq P \cup T$
$\mathcal{X}$	A characteristic function
$\mathcal{X}_{p_i}$	The characteristic function of p_i
$\mathcal{X}_S$	The characteristic function of S
$\mathcal{X}_{S_N}$	The characteristic function of S_N
$\mathcal{X}_{S_{MS}}$	The characteristic function of S_{MS}
$\mathcal{X}_{\mathcal{R}}$	The characteristic function of $\mathcal{R}$
$\mathcal{X}_{\mathcal{R}_C}$	The characteristic function of $\mathcal{R}_C$
$\mathbb{Z}$	The set of integers, $\mathbb{Z} = \{\ldots, -2, -1, 0, 1, 2, \ldots\}$
Ξ_{M,M_1}	The set of legal markings that M-equal M_1
σ	A sequence of transitions
$\vec{\sigma}$	The Parikh vector of σ
Γ	A non-oriented path
Γ_M	A non-oriented path from M_0 to M
$\vec{\Gamma}$	The counting vector of Γ
$\vec{\Gamma}_M(t)$	The algebraic sum of all occurrences of t in Γ_M
γ	A non-oriented cycle
$\vec{\gamma}(t)$	The algebraic sum of all occurrences of t in γ
δ^t	The transition function of t
ε	An encoding function
μ_i	The marking of p_i
Ω	A set of MTSIs
Ω_{p_c}	The set of MTSIs that are implemented by p_c

Index

T - #0030 - 160425 - C0 - 234/156/11 [13] - CB - 9781466577534 - Gloss Lamination